CENTRO DI RICERCA MATEMATICA
ENNIO DE GIORGI

# Nicolai V. Krylov

Probabilistic Methods of Investigating
Interior Smoothness of
Harmonic Functions
Associated with
Degenerate Elliptic Operators

SCUOLA NORMALE SUPERIORE
PISA

# Contents

# *Foreword*

In May 2004 Nicolai Krylov (University of Minnesota) visited the Centro di Ricerca Matematica Ennio De Giorgi as UMI - Lecturer with the partial support of the Unione Matematica Italiana. It is a great pleasure to see now the enlarged version of his lectures to appear in the series of Publications of the Center.

Mariano Giaquinta

# Probabilistic Methods of Investigating Interior smoothness of Harmonic Functions Associated with Degenerate Elliptic Operators

**Abstract.**  The lectures concentrate on some old and new relations between quasiderivatives of solutions to Itô stochastic equations and interior smoothness of harmonic functions associated with degenerate elliptic equations. Recent progress in the case of constant coefficients is discussed in full detail.

## Preface

We are dealing with diffusion processes $x_t \in \mathbb{R}^d$ given as solutions of the Itô equation

$$(0.0.1) \qquad dx_t = \sigma^k(x_t)\,dw_t^k + b(x_t)\,dt\,,$$

where $w_t^k$ are independent one-dimensional Brownian motions, $k = 1, \ldots, d_1$ with $d_1$ perhaps different from $d$, $\sigma^k$ and $b$ are $\mathbb{R}^d$-valued functions and the summation in $k$ is assumed. We introduce $\sigma$ as the matrix composed of the column-vectors $\sigma^k$: $\sigma = (\sigma^1, \ldots, \sigma^k)$, and define $a = (1/2)\sigma\sigma^*$,

$$Lu = a^{ij}u_{x^i x^j} + b^i u_{x^i}\,.$$

Under appropriate conditions on $\sigma$ and $b$, for any nonrandom initial data $x \in \mathbb{R}^d$, equation (0.0.1) has a unique solution $x_t(x)$. As is common we use the symbols $E_x$ and $P_x$ for the expectation of random variables and the probability of events defined in terms of $x_t(x)$ and drop the argument $x$ behind the expectation and probability signs. For a domain $D \subset \mathbb{R}^d$ we denote by $\tau = \tau(x) = \tau_D(x)$ the first exit time of $x_t(x)$ from $D$.

The work was partially supported by NSF Grant DMS-0140405

Take a smooth bounded domain $D \subset \mathbb{R}^d$, a function $g \in C^1(\bar{D})$, assume that $P_x(\tau < \infty) = 1$ for all $x \in D$ and define

$$(0.0.2) \qquad\qquad u(x) = E_x g(x_\tau).$$

The function $u$ is known as *a probabilistic solution* of the equation $Lu = 0$ in $D$ with boundary data $u = g$ on $\partial D$. However generally, $u$ does not have two derivatives needed in $L$ and the equation is understood in a generalized sense.

Considerable effort was applied to understand under which conditions $u$ is twice differentiable and does satisfy the equation. The first probabilistic results were obtained by Gikhman in [4] (1950-51) in parabolic case and Freidlin in [5] (1968) in the above setting. The techniques based on methods of the theory of partial differential equations is used in the basic sources of information about degenerate elliptic equations: Kohn-Nirenberg [6] (1967) and Oleinik-Radkevich [17] (1969). In [6] and [17] in contrast with [4] and [5] the solutions are looked for in Sobolev classes. In that framework for special domains and operators one can get additional information from Kondrat'ev [7] (1966) and Alkhutov [1] (1995).

As long as the usual second-order derivatives of $u$ are concerned the best general results are presented in Krylov [8] (1989), where even controlled processes and consequently fully nonlinear elliptic equations are considered. The results in [8] are obtained probabilistically by introducing and using *quasiderivatives* and a reduction of controlled processes in domains to controlled processes on a surface without boundary in the space having four more dimensions. Later on PDE proofs of these results appeared in Krylov [11] (1995). Specification of them for the particular case in which $L$ is the heat operator is given in Krylov [13] (2002) in multidimensional case. In Hongjie Dong [3] (submitted) necessary and sufficient conditions are found in one space dimension for the heat equation under which the solution is $k$ times continuously differentiable up to the boundary. However in these papers one assumes that $g$ is at least four times continuously differentiable. Generally, this assumption is necessary if we want to estimate the second-order derivatives of $u$ up to the boundary. Interestingly enough, in [13] and [3] where only the heat equation is treated the four dimensional extension of the domain plays a crucial role.

Naturally, the question arises as to what happens if $g$ is only once differentiable. In that case even if $L = \Delta$, one cannot assert that the first-order derivatives of $u$ are bounded up to the boundary (see, for instance, Example 4.2.1) and one only can hope to prove that inside $D$ the derivatives of $u$ exist. The fact that under various conditions on the process the first-order derivatives of $u$ can be indeed estimated inside $D$ was proved in Krylov [9] (1992) and Krylov [10] (1993). The trouble with these various conditions is that each particular case was treated by its own method. Furthermore, if

$$(0.0.3) \qquad\qquad d = 2, \quad D = \{|x| < 1\}, \quad Lu = u_{x^1 x^1},$$

the methods of [9] and [10] allowed estimating $u_{x^2}$ only for $x^1 = 0$. This is particularly disturbing because in that case one can find $u$ explicitly and then the estimates are straightforward. This unsatisfactory situation is further discussed in Remark 4.4.2.

In the last chapter of these lectures we present a unified method based on quasiderivatives which allows us to treat all cases of *constant* $\sigma$ and $b$ simultaneously. There is certain hope that our methods can be applied to variable $\sigma$ and $b$ (cf. Remark 5.2.2). However, it is unlikely that controlled processes can be treated in the same way.

Another strong motivation to come back to investigating quasiderivatives arises from the author's inability to prove Hörmander's theorem about hypoelliptic equations. The point is that, although in very many cases of fully nonlinear degenerate elliptic equations in domains using quasiderivatives played a crucial role in investigating their counterparts on manifolds of four dimensions more and then yielded *sharp* results for the original equations in domains, yet the author does not know how to use quasiderivatives in order to prove that solutions of *linear* hypoelliptic equations are smooth. In this regard limitations of known quasiderivatives are described in Section 4.5.

The lectures are organized as follows. In Chapter 1 we discuss a purely analytic method of getting interior gradient estimates for equations with *constant* coefficients in smooth strictly convex domains when $b = 0$. The case that constant $b \neq 0$ is treated by probabilistic methods in Section 5.4. At this moment we do not have analytic methods allowing us to treat the case $b \neq 0$. As has been mentioned already probabilistic methods are based on quasiderivatives which we introduce in Section 3.1. Section 3.2 contains known examples of quasiderivatives related to random time change, change of measure, and change of the driving Wiener processes. Two applications of time-change related quasiderivatives are given in Section 4.1: in the first example we assume (0.0.3) and show how to estimate $u_{x^2}$ on the $x^2$-axis, the second example shows the interior estimate of time derivative for the Cauchy problem for degenerate parabolic equations.

Sections 3.1-3.4 have much in common with Krylov [10] (1993). The same goes for Subsection 4.4.1. Part of the lectures has been taken from [9] although we provide missing details and to make the presentation more available to the reader do not try to get the results in their utmost generality.

In Section 4.3 we come back to assumption (0.0.3) and show that combining time-change and measure-change related quasiderivatives allows us to get sharp interior estimates of $u_x$ everywhere. The method of Section 4.3 does not extend to more general situations and Sections 5.1-5.4 are devoted to developing general methods of proving interior estimates by probabilistic methods and applying these estimates to equations with constant coefficients in strictly convex domains.

The lectures took their final form soon after my visiting Pisa where I enjoyed the hospitality of Centro di Ricerca Matematica Ennio De Giorgi, Collegio Puteano, Scuola Normale Superiore (Pisa, Italy, May-June, 2004) and it is my great pleasure to thank G. Da Prato and M. Giaquinta for making my visit possible.

In conclusion we introduce some notation. Above we have already used $C^1(\bar{D})$ for the space of bounded continuous and once continuously differentiable functions on $\bar{D}$ with finite norm given by

$$|g|_{1,D} = |g|_{0,D} + |g_x|_{0,D}, \quad |g|_{0,D} = \sup_{x \in D} |g(x)|,$$

where $g_x$ is the gradient of $g$. We use similar notation for spaces of functions with higher order derivatives. For $\alpha \in (0, 1]$ we introduce $C^{0,\alpha}(\Gamma)$ as the set of functions $g$ defined on a set $\Gamma$ with finite norm

$$|g|'_{\alpha,\Gamma} = |g|_{0,\Gamma} + [g]_{\alpha,\Gamma}, \quad [g]_{\alpha,\Gamma} = \sup_{x,y \in \Gamma} \frac{|g(x) - g(y)|}{|x - y|^\alpha}.$$

We use the summation convention over repeated indices and for $\xi \in \mathbb{R}^d$ set

$$u_{(\xi)} = \xi^i u_{x^i}.$$

The reader is referred to an index of these and other notation placed in the end of the lectures.

## 1. Prolog: A PDE approach to the equation $a^{ij}u_{x^ix^j} = 0$ with constant $a_{ij}$ in strictly convex domains

Let $D \in C^3$ be a uniformly convex domain in $\mathbb{R}^d$. We consider the problem

(1.0.1)        $Lu := a^{ij}u_{x^ix^j} = 0 \quad \text{in} \quad D, \quad u = g \quad \text{on} \quad \partial D.$

We assume that $a^{ij}$ are constant,

$$a = a^* \geq 0, \quad \operatorname{tr} a = 1.$$

We can always solve (1.0.1) on the hyperplanes where $a$ is nondegenerate and therefore the existence of solution $u$ of (1.0.1) is no problem for any bounded continuous $g$.

　　To state our main result we observe that there exists a concave function $\psi \in C^3(\bar{D})$ such that $\psi > 0$ in $D$, $\psi = 0$ on $\partial D$, and

$$\psi_{x^ix^j} l^i l^j \leq -1$$

in $D$ for any unit $l \in \mathbb{R}^d$. Notice that the diameter of $D$ can be easily estimated through $|\psi|_{2,D}$.

　　The goal of this section is to give a PDE proof of the following result.

THEOREM 1.0.1. *Let $g \in C^{0,1}(\bar{D})$. Then $u \in C^{0,1}_{loc}(D)$ and there is a constant $N$ depending only on $|\psi|_{3,D}$ and $d$ such that*

(1.0.2) $$|u_x| \le N\psi^{-5/2}|g|'_{1,D}.$$

REMARK 1.0.2. Take $d = 2$ and let $(x, y)$ be a generic point in $\mathbb{R}^2$. Set

$$Lu(x, y) = u_{xx}(x, y)$$

and take $D = \{(x, y) : |x|^2 + |y|^2 < 1\}$. Let $g(x)$ be Lipschitz continuous in $\mathbb{R}$ and continuously differentiable in $\mathbb{R} \setminus \{0\}$. Then the solution of problem (1.0.1) is easily shown to be

$$u(x, y) = \frac{x}{\sqrt{1 - |y|^2}} f\left(\sqrt{1 - |y|^2}\right) + h\left(\sqrt{1 - |y|^2}\right),$$

where

$$f(x) = \frac{1}{2}[g(x) - g(-x)], \quad h(x) = \frac{1}{2}[g(x) + g(-x)].$$

In particular, for $x = 0$

$$u_y(0, y) = -\frac{yh'\left(\sqrt{1 - |y|^2}\right)}{\sqrt{1 - |y|^2}}.$$

We see that, if $h'(0) > 0$, then the derivative in $y$ of $u(x, y)$ at $(0, y)$ behaves as $(0, y)$ goes to the south pole as the inverse to the square root of the distance of $(0, y)$ to the boundary of $D$.

REMARK 1.0.3. Even if $L = \Delta$ one cannot estimate the gradient of $u$ up to the boundary in terms of $|g|_{1,D}$. We show that in Example 4.2.1.

Now we start proving Theorem 1.0.1. Having in mind obvious approximations we convince ourselves that without losing generality we may assume that $a$ is nondegenerate and $D$ and $g$ are infinitely differentiable. Then $u$ is also infinitely differentiable in $\bar{D}$ and admits the representation

(1.0.3) $$u(x) = \int_{\partial D} g(y) \pi(x, dy),$$

where $\pi(x, \cdot)$ is an $L$-harmonic measure. It is well known, but not used below, that this measure has a smooth density with respect to the surface measure on $\partial D$.

By differentiating (1.0.1) in the direction of a $\xi \in \mathbb{R}^d$ and using (1.0.3) we get

(1.0.4) $$u_{(\xi)}(x) = \int_{\partial D} u_{(\xi)}(y) \pi(x, dy).$$

If we knew that $u_{(\xi)}(y)$ could be replaced with $u_{(\xi(y))}(y)$, where $\xi(y)$ is tangential to $\partial D$ at $y \in \partial D$, then upon noticing that

$$u_{(\xi(y))}(y) = g_{(\xi(y))}(y)$$

for such $\xi(y)$, we would see why Theorem 1.0.1 is true.

Now one can imagine that the idea how to make $\xi$ to be tangent to the boundary came from observing the following fact.

LEMMA 1.0.4. *For any $x \in D$ we have*

(1.0.5) $$\int_{\partial D} u_{(y-x)}(y)\,\pi(x, dy) = \int_{\partial D} u_{x^i}(y)(y^i - x^i)\,\pi(x, dy) = 0.$$

PROOF. Fix an $x_0 \in D$ and introduce

$$v(x) = u_{x^i}(x)(x^i - x_0^i).$$

It is easy to check that $Lv = 0$. Since $v(x_0) = 0$ we get (1.0.5) with $x_0$ in place of $x$ from (1.0.3) thus proving the lemma. □

Few times during my pedagogical life I observed the students' attempts to do something with an integral like

$$\int_D f(x)g(x)\,dx$$

by pulling $f$ out, then doing something with the integral of $g$ and then bringing $f$ back inside the integral. Inspired by this "clever" trick let us multiply the left-hand side of (1.0.5) by a function $p(y)$ put it inside the integral and add the result to (1.0.4). Then we get

(1.0.6) $$u_{(\xi)}(x) = \int_{\partial D} u_{(\xi(y))}(y)\,\pi(x, dy),$$

where

$$\xi(y) = \xi + p(y)(y - x).$$

Here $\xi$ and $x$ are fixed and $p(y)$ is kind of arbitrary. Since $D$ is strictly convex it is very easy to visualize how $p$ should be taken in order to make $\xi(y)$ tangent to $\partial D$ at $y$. After such choice it would only remain to replace $u_{(\xi(y))}(y)$ with $g_{(\xi(y))}(y)$.

However, (1.0.6) is certainly not true. Actually, the right-hand side of (1.0.6) is

$$\int_{\partial D} u_{(\xi)}(y)\,\pi(x, dy) + \int_{\partial D} p(y)u_{(y-x)}(y)\,\pi(x, dy),$$

where the second integral equals

$$\int_{\partial D} (pu)_{(y-x)}(y)\,\pi(x, dy) - \int_{\partial D} g(y)p_{(y-x)}(y)\,\pi(x, dy).$$

Hence by virtue of (1.0.4) instead of (1.0.6) the following is true

(1.0.7)
$$u_{(\xi)}(x) = \int_{\partial D} u_{(\xi(y))}(y)\,\pi(x, dy) + \int_{\partial D} g(y)p_{(y-x)}(y)\,\pi(x, dy)$$
$$- \int_{\partial D} (pu)_{(y-x)}(y)\,\pi(x, dy).$$

Under the above described choice of $p$ we know how to control the first two terms on the right of (1.0.7). Formula (1.0.5) suggests a way to deal with the last term as follows. Introduce

$$v(x) = \int_{\partial D} pg(y)\,\pi(x, dy).$$

Then

$$\int_{\partial D} v_{(y-x)}(y)\,\pi(x, dy) = 0.$$

We want to compare the integral on the left with the last term on the right in (1.0.7). Notice that

$$v = pg = pu \quad \text{on} \quad \partial D.$$

Therefore the tangential derivatives of $v$ and $pu$ coincide and the only difference comes from the normal derivatives of these functions. It turns out that this difference can be estimated despite the fact that each of the terms *cannot* be controlled in terms of $|g|_{1,D}$ even if $L = \Delta$ (see Remark 1.0.3).

LEMMA 1.0.5. *Let* $|p|_{2,D} < \infty$. *Then there is a constant* $N$ *depending only on* $d$ *such that on* $\partial D$ *for any* $\eta \in \mathbb{R}^d$

$$(1.0.8) \qquad |v_{(\eta)} - (pu)_{(\eta)}| \le N|\psi_x|_{0,\partial D}|p|_{2,D}|g|_{1,D}|(n, \eta)|,$$

*where* $n$ *is the unit inward normal on* $\partial D$.

PROOF. As we have pointed out before the lemma it suffices to consider only the case in which $\eta = n$. Fix a $y_0 \in \partial D$ take a small $\varepsilon > 0$ and for $x = y_0 - \varepsilon n(y_0)$ write

$$v(x) = \int_{\partial D} pg(y)\,\pi(x, dy) = p(x)\int_{\partial D} g(y)\,\pi(x, dy)$$
$$+ \int_{\partial D} [p(y) - p(x)]g(y)\,\pi(x, dy),$$

which by definition of $u$ and the fact that $v = pu$ on $\partial D$ implies that

$$[v(x) - v(y_0)] - [pu(x) - pu(y_0)] = \int_{\partial D} [p(y) - p(x)]g(y)\,\pi(x, dy) =: I.$$

By using Taylor's formula we see that

$$I \le p_{x^i}(x)\int_{\partial D} (y^i - x^i)g(y)\,\pi(x, dy)$$
$$+ N|p|_{2,D}|g|_{0,D}\int_{\partial D} |y - x|^2\,\pi(x, dy) =: p_{x^i}(x)J_i + N|p|_{2,D}|g|_{0,D}J.$$

Upon applying (1.0.5) to the function $u(x) = x^i$ we find that for each $i$

$$\int_{\partial D} (y^i - x^i)\, \pi(x, dy) = 0.$$

Hence,

$$|J_i| = \left| \int_{\partial D} (y^i - x^i)[g(y) - g(x)]\, \pi(x, dy) \right| \le |g|_{1,D} J.$$

Finally, observe that

$$J = \int_{\partial D} |y|^2 \pi(x, dy) - 2x^i \int_{\partial D} (y^i - x^i)\, \pi(x, dy) - |x|^2$$

$$= \int_{\partial D} |y|^2 \pi(x, dy) - |x|^2 = w(x),$$

where $w$ is the solution of the equation $Lw = -2\,\mathrm{Trace}\,a$ in $D$ with zero boundary data. By the maximum principle $|w| \le 2\psi$ and by combining our estimates we get

$$[v(x) - v(y_0)] - [pu(x) - pu(y_0)] \le N\psi(x)|p|_{2,D}|g|_{1,D}.$$

If we divide through this equation by $\varepsilon$ and let $\varepsilon \downarrow 0$, then we get that

$$v_{(n)} - (pu)_{(n)}$$

is less than the right-hand side of (1.0.8). Changing $g$ to $-g$ yields the estimate of $v_{(n)} - (pu)_{(n)}$ from below. The lemma is proved.                    □

We now go back to (1.0.7) and obtain

(1.0.9)
$$\left| \int_{\partial D} (pu)_{(y-x)}(y)\, \pi(x, dy) \right| = \left| \int_{\partial D} [(pu)_{(y-x)}(y) - v_{(y-x)}(y)]\, \pi(x, dy) \right|$$

$$\le N|\psi_x|_{0,\partial D}|p|_{2,D}|g|_{1,D} \int_{\partial D} |(n(y), y - x))|\, \pi(x, dy),$$

where by Hölder's inequality

$$\int_{\partial D} |(n(y), y - x))|\, \pi(x, dy) \le \sqrt{J} \le \sqrt{2\psi(x)}$$

with $J$ introduced in the proof of Lemma 1.0.5.

We also estimate the first two terms on the right in (1.0.7) assuming that $\xi(y)$ is tangent to $\partial D$ at $y$. Then

$$\left| \int_{\partial D} u_{(\xi(y))}(y)\,\pi(x,dy) \right| \leq |g|_{1,D} \int_{\partial D} |\xi(y)|\,\pi(x,dy)$$

$$\leq |g|_{1,D} \int_{\partial D} (|\xi| + |p(y)|\,|y-x|)\,\pi(x,dy)$$

$$\leq |g|_{1,D}|\xi| + |p|_{0,D}\sqrt{J}$$

$$\leq |g|_{1,D}\left(|\xi| + |p|_{0,D}\sqrt{2\psi(x)}\right),$$

$$\left| \int_{\partial D} g(y)p_{(y-x)}(y)\,\pi(x,dy) \right| \leq |g|_{0,D}|p|_{1,D}\sqrt{J} \leq |g|_{0,D}|p|_{1,D}\sqrt{2\psi(x)}.$$

Thus, owing to (1.0.7)

(1.0.10)
$$|u_{(\xi)}(x)| \leq |g|_{1,D}\left(|\xi| + |p|_{0,D}\sqrt{2\psi(x)}\right)$$
$$+ |g|_{0,D}|p|_{1,D}\sqrt{2\psi(x)} + N\sqrt{\psi(x)}|\psi_x|_{0,\partial D}|p|_{2,D}|g|_{1,D}.$$

Now it is time to choose $p(y)$. We need

$$\xi + p(y)(y-x)$$

to be tangent to $\partial D$ at $y$, that is

$$\psi_{(\xi+p(y)(y-x))}(y) = 0, \quad \psi_{(\xi)}(y) + p(y)\psi_{(y-x)}(y) = 0, \quad p(y) = -\frac{\psi_{(\xi)}(y)}{\psi_{(y-x)}(y)}$$

for $y \in \partial D$. By the way, since $D$ is convex and $\psi$ is concave, for $y \in \partial D$, we have

(1.0.11)
$$\psi_{(y-x)}(y) = -\psi(x) + \frac{1}{2}\psi_{(y-x)(y-x)}(\theta) \leq -\psi(x) < 0,$$

where $\theta$ is a point between $x$ and $y$ on the straight line passing through these points. Therefore, $p$ is well defined on $\partial D$. By using for the first time our assumption that the $C^3$-norm of $\psi$ is allowed to enter our estimates and having in mind (1.0.11) we can continue $p$ from $\partial D$ inside $D$ in such a way that

$$|p|_2 \leq N|\xi|\psi^{-3}(x).$$

Then (1.0.10) yields

$$|u_{(\xi)}(x)| \leq N|g|_{1,D}|\xi|(1 + \psi^{-5/2}(x)),$$

which is equivalent to (1.0.2) since $\xi$ is arbitrary. The theorem is proved.

REMARK 1.0.6. Estimate (1.0.2) can be slightly improved. It turns out that

$$(1.0.12) \qquad |u_{(\xi)}| \le N|\xi|\psi^{-2},$$

where $N$ depends only on $|\psi|_{3,D}$ and $d$.

To see that take a smooth strictly positive function $q(x)$ defined on $\bar{D}$, let $r = q^{-1}$ and

$$\bar{u}(x) := \int_{\partial D} rg(y)\,\pi(x, dy).$$

Notice that by Lemma 1.0.5 with $u$, $q$, and $\bar{u}$ in place of $v$, $p$, and $u$, respectively, we have on $\partial D$ that

$$u_{(\xi)}(y) \le (q\bar{u})_{(\xi)}(y) + N|q|_{2,D}|rg|_{1,D}|\xi| = \bar{u}_{(q(y)\xi)}(y) + rgq_{(\xi)}(y)$$
$$+ N|q|_{2,D}|rg|_{1,D}|\xi| \le \bar{u}_{(q(y)\xi)}(y) + N|q|_{2,D}|r|_{1,D}|g|_{1,D}|\xi|.$$

Hence,

$$(1.0.13) \qquad u_{(\xi)}(x) \le \int_{\partial D} \bar{u}_{(q(y)\xi)}(y)\,\pi(x, dy) + N|q|_{2,D}|r|_{1,D}|g|_{1,D}|\xi|.$$

Also owing to (1.0.9)

$$\left| \int_{\partial D} p\bar{u}_{(y-x)}\,\pi(x, dy) \right| \le \left| \int_{\partial D} \bar{u}\,p_{(y-x)}\,\pi(x, dy) \right|$$
$$+ N|p|_{2,D}|rg|_{1,D}\sqrt{\psi(x)} \le N|p|_{2,D}|r|_{1,D}|g|_{1,D}.$$

Therefore, (1.0.13) implies that

$$(1.0.14) \quad u_{(\xi)}(x) \le \int_{\partial D} \bar{u}_{(\bar{\xi}(y))}(y)\,\pi(x, dy) + N|r|_{1,D}|g|_{1,D}(|q|_{2,D}|\xi| + |p|_{2,D}),$$

where

$$\bar{\xi}(y) = q(y)\xi + p(y)(y - x).$$

If we make $\bar{\xi}(y)$ to be tangential to $\partial D$ at $y$, then

$$\bar{u}_{(\bar{\xi}(y))}(y) = (rg)_{(\bar{\xi}(y))}(y) \le N|r|_{1,D}|g|_{1,D}(|q|_{0,D}|\xi| + |p|_{0,D})$$

and (1.0.14) yields

$$(1.0.15) \qquad u_{(\xi)}(x) \le N|r|_{1,D}|g|_{1,D}(|q|_{2,D}|\xi| + |p|_{2,D}).$$

To do the right choice of $q, p$ we have to satisfy

$$q\psi_{(\xi)}(y) + p\psi_{(y-x)}(y) = 0$$

for $y \in \partial D$. On $\partial D$ we take

$$q(y) = -\psi_{(y-x)}(y), \quad p(y) = \psi_{(\xi)}(y)$$

and continue $q, p$ preserving their smoothness and the inequality

$$q(y) \geq (1/2)\psi(x)$$

(cf. (1.0.11)). Then

$$|r|_{1,D} \leq N\psi^{-2}(x)$$

and (1.0.15) yields

$$u_{(\xi)} \leq N|\xi|\psi^{-2}$$

for all $\xi$. The arbitrariness of $\xi$ proves (1.0.12).

In the author's opinion even the better rate $\psi^{-2}$ does not look right. However we only have the following.

CONJECTURE. It holds that

$$|u_{(\xi)}| \leq N|g|_{1,D}(|\xi| + |\psi_{(\xi)}|\psi^{-1/2}).$$


## 2. Itô stochastic equations

The purpose of this chapter is to remind the reader some basic facts from the theory of stochastic equations especially related to differentiability of their solutions with respect to parameters and random time change. The exposition here mostly follows [9] and [12].

We use the following setting throughout the chapter. Let $(\Omega, \mathcal{F}, P)$ be a complete probability space and let $(w_t, \mathcal{F}_t)$ be a $d_1$-dimensional Wiener process on this space defined for $t \in [0, \infty)$, with $\sigma$-algebras $\mathcal{F}_t$ being complete with respect to $\mathcal{F}, P$.


### 2.1. Solvability under monotonicity conditions

Here we present the most general existence theorem for general Itô stochastic equations.

Assume that, for any $\omega \in \Omega$, $t \geq 0$, and $x \in \mathbb{R}^d$, we are given a $d \times d_1$-dimensional matrix $\sigma(t, x)$ and a $d$-vector $b(t, x)$ (as usual we suppress $\omega$ in the arguments of random variables). We assume that $\sigma$ and $b$ are continuous in $x$ for any $\omega, t$, measurable in $(\omega, t)$ for any $x$, and $\mathcal{F}_t$-measurable in $\omega$ for any $t$ and $x$.

For a matrix $\sigma$ write

$$\|\sigma\|^2 := \sum_{ij} |\sigma^{ij}|^2$$

and assume that, for any finite $T$ and $R$ and $\omega \in \Omega$, we have

(2.1.1)                     $$\int_0^T \sup_{|x| \le R} \{\|\sigma(t, x)\|^2 + |b(t, x)|\} \, dt < \infty.$$

REMARK 2.1.1. It is shown in [15] that, under the monotonicity condition (2.1.2) and the coercivity condition (2.1.3) below, condition (2.1.1) is satisfied if

$$\int_0^T \{\|\sigma(t, x)\|^2 + |b(t, x)|\} \, dt < \infty \quad \forall x \in \mathbb{R}^d.$$

THEOREM 2.1.2. *In addition to the above assumptions, for all $R$, $t \in [0, \infty)$, $|x|, |y| \le R$, and $\omega$, let*

(2.1.2) $2(x - y, b(t, x) - b(t, y)) + \|\sigma(t, x) - \sigma(t, y)\|^2 \le K_t(R)|x - y|^2$,

(2.1.3)                     $2(x, b(t, x)) + \|\sigma(t, x)\|^2 \le K_t(1)(1 + |x|^2)$,

*where $K_t(R)$ are certain $\mathcal{F}_t$-adapted nonnegative processes satisfying*

$$\alpha_T(R) := \int_0^T K_t(R) \, dt < \infty$$

*for all $R, T \in [0, \infty)$, and $\omega$. Also let $x_0$ be an $\mathcal{F}_0$-measurable d-dimensional vector. Then the Itô equation*

(2.1.4)                     $$dx_t = \sigma(t, x_t) \, dw_t + b(t, x_t) \, dt, \quad t \ge 0$$

*with initial condition $x_0$ has a solution, which is, moreover, unique up to indistinguishability.*

REMARK 2.1.3. It follows from inequalities like

$$|b(t, x)| \le |b(t, 0)| + |b(t, x) - b(t, 0)|$$

that assumptions (2.1.1), (2.1.2), and (2.1.3) are satisfied if

$$\int_0^T \left[ \|\sigma(t, 0)\|^2 + |b(t, 0)| \right] dt < \infty$$

and the Lipschitz condition

$$\|\sigma(t, x) - \sigma(t, y)\| + |b(t, x) - b(t, y)| \le K|x - y|$$

holds for all $t \in [0, T]$ and $x, y \in \mathbb{R}^d$, where $K$ is a constant. However (2.1.2) and (2.1.3) are also satisfied if, for instance, $d = 1$, $\sigma \equiv 0$, and $b(t, x)$ is *any* decreasing function of $x$ equal zero at $x = 0$. The one dimensional equation

$$x_t = - \int_0^t x_s^3 \, ds + w_t \,,$$

that according to Theorem 2.1.2 has a unique solution defined for all time, can be a good illustration of the theorem.

REMARK 2.1.4. Usually one proves the solvability of Itô stochastic equations not only imposing the Lipschitz condition but also assuming that

$$E|x_0|^2 < \infty \,.$$

Before proving the theorem we derive three lemmas.

LEMMA 2.1.5. *Let $y_t$ be a continuous nonnegative $\mathcal{F}_t$-adapted process, $\gamma$ a finite stopping time, $N$ a constant, and assume $E y_\tau \leq N$ for any stopping time $\tau \leq \gamma$. Then*

$$P\{\sup_{t \leq \gamma} y_t \geq \varepsilon\} \leq N/\varepsilon$$

*for any $\varepsilon > 0$.*

For the proof, it suffices to take $\tau = \gamma \wedge \inf\{t \geq 0 : y_t \geq \varepsilon\}$, to notice that $\tau \leq \gamma$ and

$$\{\omega : \sup_{t \leq \gamma} y_t \geq \varepsilon\} = \{\omega : y_\tau \geq \varepsilon\} \,,$$

and to use Chebyshev's inequality.

LEMMA 2.1.6. *Let $\xi_t, \eta_t, m_t$ be continuous real-valued processes on $[0, \infty)$. Assume that $\xi_t \geq 0$, $\eta_t \geq 0$, $\eta_t$ is nondecreasing and $m_t$ is a local $\mathcal{F}_t$-martingale starting at zero. Finally, let*

$$(2.1.5) \qquad\qquad \xi_t \leq \eta_t + m_t$$

*for all $t \geq 0$. Then for any finite stopping time $\tau$ we have*

$$(2.1.6) \qquad\qquad E\xi_\tau \leq E\eta_\tau \,.$$

If $\tau$ is a localizing stopping time for $m_t$, equation (2.1.6) follows from (2.1.5) by definition. In the general case it suffices to take localizing times $\tau_n \to \infty$, use (2.1.6) with $\tau \wedge \tau_n$ in place of $\tau$ then let $n \to \infty$ and use Fatou's lemma on the left and the inequality $\eta_{\tau \wedge \tau_n} \leq \eta_\tau$ on the right.

In the following lemma the continuity of $\sigma$ and $b$ in $x$ is not needed.

LEMMA 2.1.7. *Suppose that, for $n = 1, 2, \ldots$, we are given continuous $d$-dimensional $\mathcal{F}_t$-adapted processes $x_t^n$ on $\Omega \times [0, \infty)$ such that $x_0^n = x_0$ and*

$$dx_t^n = \sigma(t, x_t^n + p_t^n)\, dw_t + b(t, x_t^n + p_t^n)\, dt \quad t \geq 0$$

*for some $\mathcal{F}_t$-adapted processes $p_t^n$ which are measurable in $(\omega, t)$. For $n = 1, 2, \ldots$ and $R \geq 0$, let $\tau^n(R)$ be stopping times such that*

(i) *"before $\tau^n(R)$ the processes $x_t^n$ and $p_t^n$ are under control":*

(2.1.7)                     $|x_t^n| + |p_t^n| \leq R \quad for \quad 0 < t \leq \tau^n(R)$,

(ii) *"$p_t^n$ tends to zero on $(0, \tau^n(R))$":*

(2.1.8)                $\lim_{n \to \infty} E \int_0^{T \wedge \tau^n(R)} |p_t^n|\, dt = 0 \quad \forall R, T \in [0, \infty)$,

(iii) *there exists a nonrandom function $r(R)$ such that*

$$r(R) \to \infty \quad as \quad R \to \infty$$

*and "$\tau^n(R)$ is bigger than the first exit time of $x_t^n$ from the ball of radius $r(R)$ centered at the origin":*

(2.1.9)     $\varlimsup_{R \to \infty} \varlimsup_{n \to \infty} P\{\tau^n(R) \leq T, \sup_{t \leq \tau^n(R)} |x_t^n| \leq r(R)\} = 0 \quad \forall T \in [0, \infty)$.

*Then, for any $T \in [0, \infty)$, we have*

(2.1.10)                              $\sup_{t \leq T} |x_t^n - x_t^m| \overset{P}{\to} 0$

*as $n, m \to \infty$.*

REMARK 2.1.8. Observe that in (2.1.7) the value $t = 0$ is excluded, so that if for instance $\tau^n(\omega, R) = 0$ and $|x_0^n(\omega)| > R$, condition (2.1.7) is satisfied at that $\omega$. Generally, the smaller $\tau^n(R)$ are, the less restrictive conditions (2.1.7) and (2.1.8) are. However owing to (2.1.10) one cannot take $\tau^n(R)$ too small.

An example that assumptions (i)-(iii) are satisfied is obtained if

$$|p_t^n| \leq R/n \quad for \quad 0 < t \leq \tau^n(R) := \inf\{t \geq 0 : |x_t^n| \geq cR\},$$

where $c > 0$ is a constant.

PROOF OF LEMMA 2.1.7. Without loss of generality (see (2.1.1)), we assume that

$$|b(t, x)| \leq K_t(R) \quad for \quad |x| \leq R.$$

Furthermore, it is easy to find stopping times $\tau(R)$ such that $\tau(R) \overset{P}{\to} \infty$ as $R \to \infty$ and $\alpha_{t \wedge \tau(R)}(R)$ is a bounded process for any $R > 0$. For instance, it suffices to take

$$\tau(R, u) := \inf\{t \geq 0 : \alpha_t(R) \geq u\},$$

to observe that, obviously,

$$P\{\tau(R, u) \leq R\} \to 0 \quad \text{as} \quad u \to \infty,$$

to find $u(R)$ so that

$$P\{\tau(R, u(R)) \leq R\} \leq 1/R,$$

and to set $\tau(R) = \tau(R, u(R))$. For such $\tau(R)$, the transition from $\tau^n(R)$ to $\tau^n(R) \wedge \tau(R)$ preserves all the assumptions of the lemma and, therefore, without loss of generality, we may and will assume that

$$\alpha_t(R) \leq u(R) \quad \text{for} \quad t \leq \tau^n(R),$$

where $u(R)$ are some finite constants.

Observe that now

(2.1.11)
$$\lim_{n \to \infty} E \int_0^{T \wedge \tau^n(R)} |p_t^n| K_t(R) \, dt = 0 \quad \forall R, T \in [0, \infty).$$

Indeed, to see this, we merely split the integrand in (2.1.11) into a sum using

$$1 = I_{K_t(R) \geq i} + I_{K_t(R) < i}$$

and then, using (2.1.8) and the inequalities $\alpha_t(R) \leq u(R)$ and $|p_t^n| \leq R$ for $t \leq \tau^n(R)$, we let first $n \to \infty$ and then $i \to \infty$, relying upon the dominated convergence theorem in the second passage to the limit.

Next, fix $n, m$, and $R$ temporarily and set

$$x_t = x_t^n, \quad y_t = x_t^m, \quad p_t = p_t^n, \quad q_t = p_t^m,$$
$$\psi_t = \psi_t(R) = \exp(-2\alpha_t(R) - |x_0|).$$

According to Itô's formula,

$$d(|x_t - y_t|^2 \psi_t) = \psi_t[2(x_t - y_t, b(t, x_t + p_t) - b(t, y_t + q_t))$$
$$+ \|\sigma(t, x_t + p_t) - \sigma(t, y_t + q_t)\|^2 - 2K_t(R)|x_t - y_t|^2] \, dt + d\beta_t,$$

where $\beta_t = \beta_t^{nm}$ is certain local martingale with $\beta_0 = 0$. To transform the right-hand side, we write

$$x_t - y_t = (x_t + p_t) - (y_t + q_t) - p_t + q_t.$$

Then we use the monotonicity condition (2.1.2) and the inequalities

$$2|x_t - y_t|^2 \geq |(x_t + p_t) - (y_t + q_t)|^2 - 2|p_t - q_t|^2, \quad |p_t - q_t|^2 \leq 2R(|p_t| + |q_t|),$$

the last of which holds at least for

$$0 < t \leq \gamma^{nm}(R) := \tau^n(R) \wedge \tau^m(R).$$

Finally, we also use the inequality $|b(t, x)| \leq K_t(R)$ for $|x| \leq R$ and find that

$$(2.1.12) \qquad |x_t^n - x_t^m|^2 \psi_t(R) \leq \lambda_t^n(R) + \lambda_t^m(R) + \beta_t^{nm}(R), \quad t \leq \gamma^{nm}(R),$$

where

$$\lambda_t^k(R) := 4(R + 1) \int_0^t |p_s^k| K_s(R) \, ds.$$

From (2.1.12) it follows by Lemma 2.1.6 that, for any $R, T \in [0, \infty)$ and any stopping time $\tau \leq T \wedge \gamma^{nm}(R)$, we have

$$(2.1.13) \qquad E|x_\tau^n - x_\tau^m|^2 \psi_\tau(R) \leq E\lambda_{T \wedge \tau^n(R)}^n(R) + E\lambda_{T \wedge \tau^m(R)}^m(R).$$

Owing to (2.1.11), the right-hand side in (2.1.13) tends to zero, and since $\tau$ is arbitrary, by Lemma 2.1.5 we get

$$\sup_{t \leq \gamma^{nm}(R) \wedge T} |x_t^n - x_t^m|^2 \psi_t(R) \xrightarrow{P} 0$$

as $n, m \to \infty$. The factor $\psi_t(R)$ can be dropped since it is positive and independent of $n, m$. We now see that, to prove (2.1.10), it only remains to show that

$$(2.1.14) \qquad \lim_{R \to \infty} \overline{\lim_{n \to \infty}} P\{\tau^n(R) \leq T\} = 0 \quad \forall T \in [0, \infty).$$

So far we have not used the growth condition (2.1.3). By using this condition and assuming $K_t(R) \geq K_t(1)$ for $R \geq 1$, which does not restrict generality, similarly to (2.1.12) we get that, for $t \leq \tau^n(R)$,

$$(1 + |x_t^n|^2)\psi_t(1) \leq (1 + |x_0|^2)e^{-|x_0|} + \lambda_t^n(R) + \beta_t^n,$$

where $\beta_t$ is a local martingale starting at zero. Then similarly to (2.1.13) we obtain that, for any stopping time $\tau \leq T \wedge \tau^n(R)$,

$$E|x_\tau^n|^2 \psi_\tau(1) \leq N + E\lambda_{T \wedge \tau^n(R)}^n(R),$$

where $N$ is the least upper bound of $(1 + |x|^2)e^{-|x|}$. From the above stated properties of $\lambda_t^n(R)$ and from Lemma 2.1.5 we further obtain that

$$\lim_{c \to \infty} \sup_{R \geq 0} \overline{\lim_{n \to \infty}} P\{ \sup_{t \leq T \wedge \tau^n(R)} |x_t^n|^2 \psi_t(1) \geq c \} = 0 \quad \forall T \in [0, \infty).$$

Again since $\psi_t(1)$ is independent of $n$ and $R$, this term can be dropped, so that

$$\lim_{c \to \infty} \sup_{R \geq 0} \overline{\lim_{n \to \infty}} P\{ \sup_{t \leq T \wedge \tau^n(R)} |x_t^n|^2 \geq c\} = 0 \quad \forall T \in [0, \infty).$$

Since $r(R) \to \infty$ as $R \to \infty$, it follows that, for all $T \in [0, \infty)$

$$\lim_{R \to \infty} \overline{\lim_{n \to \infty}} P\{ \sup_{t \leq \tau^n(R)} |x_t^n| \geq r(R), \tau^n(R) \leq T\}$$

$$\leq \lim_{c \to \infty} \sup_{R \geq 0} \overline{\lim_{n \to \infty}} P\{ \sup_{t \leq \tau^n(R)} |x_t^n| \geq c, \tau^n(R) \leq T\} = 0.$$

Combining this with the assumption (2.1.9) we get (2.1.14). The lemma is proved.

PROOF OF THEOREM 2.1.2. We are going to use Euler's method. Define the processes $x_t^n$ so that $x_0^n = x_0$ and

$$dx_t^n = \sigma(t, x_{k/n}^n) \, dw_t + b(t, x_{k/n}^n) \, dt$$

for $t \in [k/n, (k+1)/n]$. Clearly $x_t^n$ satisfy

(2.1.15)
$$dx_t^n = \sigma(t, x_{\kappa(n,t)}^n) \, dw_t + b(t, x_{\kappa(n,t)}^n) \, dt,$$
$$dx_t^n = \sigma(t, x_t^n + p_t^n) \, dw_t + b(t, x_t^n + p_t^n) \, dt,$$

where

$$\kappa(n, t) = [tn]/n, \quad p_t^n := x_{\kappa(n,t)}^n - x_t^n.$$

Also introduce

$$\tau^n(R) = \inf\{t \geq 0 : |x_t^n| \geq R/3\}, \quad r(R) = R/4.$$

Then, we have

$$|p_t^n| \leq 2R/3, \quad |x_t^n| \leq R/3 \quad \text{for} \quad 0 < t \leq \tau^n(R).$$

(Observe that $t = 0$ is excluded and has to be excluded since for $|x_0| > R/3$ we have $\tau^n(R) = 0$ and for $t \in [0, \tau^n(R)]$ it holds that $|x_t^n| = |x_0| > R/3$.) In particular, the event entering (2.1.9) is empty and this condition is satisfied. Moreover,

$$-p_t^n = \int_{\kappa(n,t)}^t \sigma(s, x_{\kappa(n,t)}^n) \, dw_s + \int_{\kappa(n,t)}^t b(s, x_{\kappa(n,t)}^n) \, ds.$$

Hence according to well known inequalities for stochastic integrals, for any $\varepsilon, \delta > 0$, we have

(2.1.16)
$$P\{|p_t^n| \geq 2\varepsilon, t \leq \tau^n(R)\} \leq P\left\{ \int_{\kappa(n,t)}^t \sup_{|x| \leq R} |b(s, x)| \, ds \geq \varepsilon \right\}$$
$$+ P\left\{ \int_{\kappa(n,t)}^t \sup_{|x| \leq R} \|\sigma(s, x)\|^2 \, ds \geq \delta \right\} + \delta/\varepsilon^2.$$

In view of (2.1.1), the last two probabilities go to zero as $n \to \infty$ for any $t, \varepsilon, \delta, R > 0$. This shows that the left-hand side of (2.1.16) goes to zero as $n \to \infty$. In other words, the product of $|p_t^n|$ and the indicator of $\{t \leq \tau^n(R)\}$ tends to zero in probability. Since this product is less than $2R/3$, its expectation is bounded in $t, n$ and tends to zero as $n \to \infty$. An application of Fubini's theorem and the dominated convergence theorem proves that (2.1.8) holds.

Now by Lemma 2.1.7 and the completeness of the space of processes with respect to the uniform convergence in probability, we get that there exists a continuous process $x_t$ such that

$$\sup_{t \leq T} |x_t^n - x_t| \overset{P}{\to} 0 \quad \forall T \in [0, \infty).$$

Here one may replace $t$ in $x_t^n$ and $x_t$ with $\kappa(n, t)$. Moreover, one can do this only for $x_t^n$ since $x_t$ is continuous in $t$. Then, using the continuity of $\sigma$ and $b$ in $x$ and passing to the limit in (2.1.15), we see that $x_t$ is indeed a solution of equation (2.1.4) with the initial condition $x_0$. Its uniqueness and even continuous dependence on the initial data we prove in the following theorem. The theorem is proved.

THEOREM 2.1.9. *Let the assumptions of Theorem 2.1.2 apart from assumption (2.1.3) be satisfied. Let $x_0, x_0^n$ be $\mathbb{R}^d$–valued $\mathcal{F}_0$–measurable vectors such that*

$$P - \lim_{n \to \infty} x_0^n = x_0.$$

*Assume that equation (2.1.4) has solutions $x_t$ and $x_t^n$ with initial data $x_0$ and $x_0^n$, respectively. Then*

(2.1.17)          $$P - \lim_{n \to \infty} \sup_{t \leq T} |x_t^n - x_t| = 0 \quad \forall T \in [0, \infty).$$

PROOF. Without loss of generality, we may and will assume that $x_0^n \to x_0$ almost surely. Then the process

$$\phi_t(R) := \exp(-\alpha_t(R) - \sup_n |x_0^n|)$$

satisfies $\phi_t(R) > 0$. Next, for

$$\gamma^n(R) = \inf\{t \geq 0 : |x_t^n| + |x_t| \geq R\} \wedge T$$

as in the proof of Lemma 2.1.7 we find that

$$|x_{t \wedge \gamma^n(R)}^n - x_{t \wedge \gamma^n(R)}|^2 \phi_{t \wedge \gamma^n(R)}(R) \leq |x_0^n - x_0|^2 \phi_0 + m_t^n(R),$$

where $m_t^n(R)$ is a local martingale starting at zero. Since $|x_0^n - x_0| \to 0$ in probability and, obviously, $|x_0^n - x_0|^2 \phi_0$ is bounded by a constant, we have

$$E|x_0^n - x_0|^2 \phi_0 \to 0$$

and Lemmas 2.1.5 and 2.1.6 imply that

$$P - \lim_{n\to\infty} \sup_{t\leq T} |x^n_{t\wedge\gamma^n(R)} - x_{t\wedge\gamma^n(R)}|^2 \exp(-\phi_{t\wedge\gamma^n(R)}(R)) = 0,$$

(2.1.18)

$$P - \lim_{n\to\infty} \sup_{t\leq T} |x^n_{t\wedge\gamma^n(R)} - x_{t\wedge\gamma^n(R)}| = 0$$

for any finite $T, R$. Add to this that

$$P\{\gamma^n(R) < T\} \leq P\{\sup_{t\leq T}(|x^n_{t\wedge\gamma^n(R)}| + |x_{t\wedge\gamma^n(R)}|) \geq R\}$$

$$\leq P\{\sup_{t\leq T} |x^n_{t\wedge\gamma^n(R)} - x_{t\wedge\gamma^n(R)}| \geq 1\} + P\{2\sup_{t\leq T} |x_t| \geq R - 1\}.$$

Then from (2.1.18) we get

$$\lim_{R\to\infty} \overline{\lim_{n\to\infty}} P\{\gamma^n(R) < T\} = 0,$$

which along with (2.1.18) implies (2.1.17). The theorem is proved.  □

Above we have only used Lemma 2.1.7 when the probability in (2.1.9) is zero. Later on we will need the following result in the proof of which the verification of (2.1.9) is not that trivial.

THEOREM 2.1.10. *Let the assumptions of Theorem 2.1.2 be satisfied. Let* $\sigma^n(t, x)$ *and* $b^n(t, x)$ *be some functions satisfying the conditions imposed on* $\sigma$ *and* $b$ *before Remark 2.1.1 and let processes* $y^n_t$ *be such that* $y^n_0 \to x_0$ *in probability and*

(2.1.19) $$dy^n_t = \sigma^n(t, y^n_t) \, dw_t + b^n(t, y^n_t) \, dt \quad t \geq 0.$$

*Define*

$$\gamma^n(R) = \inf\{t \geq 0 : |y^n_t| \geq R\}$$

*and assume that, for all* $R, T \in [0, \infty)$, *we have*

(2.1.20) $$\int_0^{T\wedge\gamma^n(R)} [\|\sigma^n - \sigma\|^2 + |b^n - b|](t, y^n_t) \, dt \xrightarrow{P} 0.$$

*Then* $y^n_t \to x_t$ *uniformly on* $[0, T]$ *in probability for any* $T \in [0, \infty)$, *where* $x_t$ *is the process from Theorem 2.1.2.*

PROOF. Set $x^{2n}_t = y^n_t - p^{2n}_t$, where

$$p^{2n}_t = \int_0^t (\sigma^n - \sigma)(s, y^n_s) \, dw_s + \int_0^t (b^n - b)(s, y^n_s) \, ds - y^n_0 + x_0,$$

and

$$x^{2n+1}_t = x_t, \quad p^{2n+1}_t = 0.$$

Also define

$$\tau^n(R) = \inf\{t \geq 0 : |x_t^n| + |p_t^n| \geq R\}, \quad r(R) = R/2.$$

Clearly

$$\tau^{2n}(R) \leq \gamma^n(R), \quad |p_t^n|, |y_t^n| \leq R \quad \text{for} \quad 0 < t \leq \tau^{2n}(R).$$

From this and (2.1.20) similarly to (2.1.16), we obtain that, as $n \to \infty$,

$$\mu^n(T, R) := \sup_{t \leq T} |p_t^n| I_{0 \leq t \leq \tau^n(R)} \overset{P}{\to} 0.$$

This certainly yields (2.1.8) and since the probability in (2.1.9) is less than

$$P\{\mu^n(T, R) \geq R/2\},$$

condition (2.1.9) is fulfilled as well. Other conditions of Lemma 2.1.7 are also satisfied and we conclude that

$$x_t^{2n} = y_t^n - p_t^{2n} \to x_t$$

uniformly on $[0, T]$ in probability for any $T \in [0, \infty)$. It only remains to notice that, for $R, \varepsilon > 0$, we have

$$P\{\sup_{t \leq T} |p_t^{2n}| \geq \varepsilon\} \leq P\{\mu^{2n}(T, R) \geq \varepsilon\} + P\{\tau^{2n}(R) \leq T\}$$

$$\leq 2P\{\mu^{2n}(T, R) \geq \varepsilon\} + P\{\sup_{t \leq T} |x_t^{2n}| \geq R - \varepsilon\},$$

which after letting first $n \to \infty$ and then $R \to \infty$ implies that $p_t^{2n} \to 0$ uniformly on $[0, T]$ in probability for any $T \in [0, \infty)$. The theorem is proved. $\square$

## 2.2. Smoothness of solutions with respect to a parameter

Here Theorem 2.1.10 will play a major role. We will be investigating the dependence of solutions of stochastic equations on parameters under *the assumption* that the solutions exist and do not exit from a domain in the phase space. In the previous section we saw some conditions for existence of solutions.

We need a parameter space $\mathbb{R}^k$, two domains

$$D \subset \mathbb{R}^d, \quad Q = \{q \in \mathbb{R}^k : |q| < 1\},$$

and a natural number $m \geq 1$.

Assume that, for any $\omega \in \Omega$, $t \geq 0$, $q \in Q$, and $x \in D$, we are given a $d \times d_1$-dimensional matrix $\sigma(t, x, q)$ and a $d$-vector $b(t, x, q)$. We assume that $\sigma$ and $b$ are measurable in $(\omega, t)$ for any $x, q$, $\mathcal{F}_t$-measurable in $\omega$ for any $t$,

$x, q$, and $m$ times continuously differentiable with respect to $(x, q) \in D \times Q$ for any $\omega$ and $t$.

Next take a function $x_0(q) = x_0(\omega, q)$ such that $x_0(q)$ is $\mathcal{F}_0$-measurable and for each $\omega$

$$x(\cdot) : Q \to D.$$

Assume that for any $q \in Q$ there is a continuous $\mathcal{F}_t$-adapted process $x_t(q)$ satisfying the equation

$$(2.2.1) \qquad x_t(q) = x(q) + \int_0^t \sigma(s, x_s(q), q) \, dw_s + \int_0^t b(s, x_s(q), q) \, ds$$

for all $t$ and such that $x_t(q) \in D$ for all $t$. In particular, we assume that the terms on the right in (2.2.1) make sense that is

$$(2.2.2) \qquad \int_0^T (\|\sigma(s, x_s(q), q)\|^2 + |b(s, x_s(q), q)|) \, ds < \infty$$

for any $T \in [0, \infty)$ (a.s.). Since a continuous in $t$ function $x_t(q)$ lies in $D$ up to any finite time $T$, it actually belongs to a compact subset of $D$ and the following assumption which we make is stronger than (2.2.2).

Suppose that we are given bounded domains $D(n)$ such that

$$D(n) \subset \bar{D}(n) \subset D(n+1) \subset D, \quad n = 1, 2, \ldots, \quad \bigcup_n D(n) = D$$

and for any $\omega$ and finite $T, n$

$$(2.2.3) \qquad \sum \int_0^T \sup_{x \in D(n)} \sup_{q \in Q} \{\|D^i \sigma(t, x, q)\|^2 + |D^i b(t, x, q)|\} \, dt < \infty,$$

where $D^i$ stands for arbitrary derivative of order $i$ with respect to $(x, q)$ and the summation is taken over all derivatives of all orders $i \leq m$.

If there is no stochastic term in (2.2.1), then from the theory of ODE one knows that the solutions are differentiable with respect to $q$ if this is true for the initial data. Also the partial derivatives satisfy a linear system which is obtained by formal differentiation of (2.2.1). Here we show that the same is true in the general situation as well. In particular, the first and, if $m \geq 2$, the second directional derivatives in the direction of a vector $\kappa \in \mathbb{R}^k$ of the solution of (2.2.1) satisfy

$$(2.2.4) \qquad d\xi_t = \partial(\xi_t, \kappa)\sigma(t, x_t, q) \, dw_t + \partial(\xi_t, \kappa)b(t, x_t, q) \, dt,$$

$$(2.2.5) \qquad \begin{aligned} d\eta_t = {}& [\partial(\eta_t, 0)\sigma(t, x_t, q) + \partial^2(\xi_t, \kappa)\sigma(t, x_t, q)] \, dw_t \\ & + [\partial(\eta_t, 0)b(t, x_t, q) + \partial^2(\xi_t, \kappa)\sigma(t, x_t, q)] \, dt, \end{aligned}$$

where for any function $u(x, q)$, $\xi \in \mathbb{R}^d$, and $\kappa \in \mathbb{R}^k$

$$\partial(\xi, \kappa)u(x, q) = u_{x^i}(x, q)\xi^i + u_{q^i}(x, q)\kappa^i ,$$

$$\partial^2(\xi, \kappa)u(x, q) = u_{x^i x^j}(x, q)\xi^i\xi^j + 2u_{x^i q^j}(x, q)\xi^i\kappa^j + u_{q^i q^j}(x, q)\kappa^i\kappa^j .$$

We need few definitions.

DEFINITION 2.2.1. Given a random process $y_t(r)$ defined for $t \geq 0$ and $r \in (r_1, r_2)$, where $r_1 < r_2$, and a point $r_0 \in (r_1, r_2)$, we say that $y_t(r)$ is *t-locally uniformly continuous at point* $r = r_0$ *in probability* if

$$P - \lim_{r \to r_0} \sup_{t \leq T} |y_t(r) - y_t(r_0)| = 0 \quad \forall T \in [0, \infty).$$

Also, if a random process $\zeta_t$ is defined for $t \in [0, \infty)$, we call $\zeta_t$ a *t-locally uniform derivative of* $y_t(r)$ *in probability* with respect to $r$ at $r = r_0$ if

$$P - \lim_{\varepsilon \to 0} \sup_{t \leq T} |\varepsilon^{-1}[y_t(r_0 + \varepsilon) - y_t(r_0)] - \zeta_t| = 0 \quad \forall T \in [0, \infty).$$

We write

$$\zeta_t = P - \partial y_t(r)/\partial r|_{r=r_0}$$

and notice that two different processes $\zeta_t$ and $\zeta'_t$ can satisfy this equality only if they are indistinguishable.

In an obvious way one defines *t-locally uniform continuity* (on $(r_1, r_2)$), higher order *t-locally uniform derivatives in probability* and also partial derivatives and directional derivatives if $y_t$ depends on a multidimensional parameter. In the case of processes like solutions $x_t = x_t(q)$ of (2.2.1), if $\kappa \in \mathbb{R}^k$ and $q \in Q$, we denote

$$x_{t(\kappa)}(q) = P - \partial x_t(q + r\kappa)/\partial r|_{r=0}, \quad x_{t(\kappa)(\kappa)}(q) = P - \partial x_{t(\kappa)}(q + r\kappa)/\partial r|_{r=0} .$$

Notice that it makes sense to speak about *t-locally uniform continuity* in probability of *t-locally uniform derivatives* in probability.

REMARK 2.2.2. The above definitions make sense and will be also used in the case of random vectors rather than random processes depending on $q$.

As an example let $\xi$ be a random variable having uniform distribution on $(0, 1)$. For $r \in (0, 1)$ define a random variable depending on the parameter $r$ by $x(r) = I_{\xi < r}$. Then as is easy to see

$$P - \partial x(r)/\partial r = 0.$$

This shows that one cannot construct calculus from the notion of derivatives in probability and the notion itself may look useless.

However, this notion possesses a very important property. We know that if we are given a sequence of continuous local martingales $m_t(n)$, $n = 1, 2, \ldots$, and a continuous process $m_t$ and $m_t(n) \to m_t$ *t*-locally uniformly in probability, then $m_t$ is a local martingale as well.

In the following theorem $x_t(q)$ is the solution of (2.2.1).

THEOREM 2.2.3. *Let the above assumptions be satisfied and let $x_0(q)$ have all derivatives in probability of all orders $\leq m$ in $Q$. Then the process $x_t(q)$ has all t-locally uniform derivatives in probability of all orders $\leq m$ in $Q$.*

*Also, for any $q \in Q$ and $\kappa \in \mathbb{R}^k$, the process $x_{t(\kappa)}(q)$ satisfies (2.2.4) (where $x_t = x_t(q)$) and $x_{t(\kappa)(\kappa)}(q)$ satisfies (2.2.5) if $m \geq 2$.*

*Finally, if the derivatives of $x_0(q)$ are continuous in probability, the derivatives of $x_t(q)$ are t-locally uniformly continuous in probability.*

PROOF. First of all notice that the functions $\partial(\xi, \kappa)\sigma$ and $\partial(\xi, \kappa)b$ are affine functions of $\xi$. Furthermore, for any $T, \omega, i, j$

$$\int_0^T \{\|\sigma_{x^i}(t, x_t(q), q)\|^2 + \|\sigma_{q^j}(t, x_t(q), q)\|^2$$
$$+ |b_{x^i}(t, x_t(q), q)| + |b_{q^j}(t, x_t(q), q)|\} \, dt < \infty$$

which follows from (2.2.3) and the fact that any trajectory of $x_t(q)$ on $[0, T]$ is bound to belong to one of $D(n)$. By Theorem 2.1.2 for any fixed $q \in Q$ and $\kappa \in \mathbb{R}^k$, there exists a unique solution $\xi_t$ of (2.2.4) with initial data $\xi_0 = x_{0(\kappa)}(q)$.

Next, for $r$ small enough so that $q + r\kappa \in Q$ set

$$\xi_t^{(r)} = \xi_t^{(r)}(q) = r^{-1}[x_t(q + r\kappa) - x_t(q)],$$
$$\sigma^{(r)}(t, \zeta) = r^{-1}[\sigma(t, x_t(q + r\kappa), q + r\kappa) - \sigma(t, x_t(q), q)]$$

and introduce $b^{(r)}(t, \zeta)$ similarly. Observe that, actually, $\sigma^{(r)}(t, \zeta)$ and $b^{(r)}(t, \zeta)$ are *independent* of $\zeta$. Obviously $\xi_t^{(r)}$ satisfies the following analogue of (2.1.19)

$$d\xi_t^{(r)} = \sigma^{(r)}(t, \xi_t^{(r)}) \, dw_t + b^{(r)}(t, \xi_t^{(r)}) \, dt.$$

Now bearing in mind Theorem 2.1.10, basically, we need to check that

(2.2.6)
$$\sigma^{(r)}(t, \xi_t^{(r)}) - \partial(\xi_t^{(r)}, \kappa)\sigma(t, x_t(q), q) \to 0,$$
$$b^{(r)}(t, \xi_t^{(r)}) - \partial(\xi_t^{(r)}, \kappa)b(t, x_t(q), q) \to 0,$$

and it looks like first of all we have to establish some uniform control on the size of $|\xi_t^{(r)}|$. However, condition (2.1.20) in Theorem 2.1.10 imposes restrictions on the differences in (2.2.6) *only* before

$$\gamma^{(r)}(R) = \inf \{t \geq 0 : |\xi_t^{(r)}| \geq R\}$$

and for $t < \gamma^{(r)}(R)$ automatically

(2.2.7)     $$|\xi_t^{(r)}| \leq R, \quad |x_t(q + r\kappa) - x_t(q)| \leq rR.$$

We now claim that for any $R, T$, and $\omega$

(2.2.8)     $$\int_0^{T \wedge \gamma^{(r)}(R)} \|\sigma^{(r)}(t, \xi_t^{(r)}) - \partial(\xi_t^{(r)}, \kappa)\sigma(t, x_t(q), q)\|^2 \, dt \to 0$$

as $r \to 0$.

To prove the claim we fix $R, T,$ and $\omega$ and, first, choose $n$ such that $x_t(q) \in D(n)$ for $t \in [0, T]$, then only concentrate on $r$ so small that the $rR$-neighborhood of $D(n)$ is in $D(n')$ for certain fixed $n'$. In this case the straight segment between

$$(x_t(q + r\kappa), q + r\kappa) \quad \text{and} \quad (x_t(q), q)$$

lies in $D(n') \times Q$ for any $t < \gamma^{(r)}(R)$ owing to (2.2.7). Hence again due to (2.2.7) and the mean value theorem for $t < \gamma^{(r)}(R)$

$$\|\sigma^{(r)}(t, \xi_t^{(r)})\| \le N \sup_{x \in D(n')} \sup_{q \in Q} \sum_{i,j} \left( \|\sigma_{x^i}(t, x, q)\| + \|\sigma_{q^j}(t, x, q)\| \right),$$

where $N = N(d, d_1)(r^{-1}R + |\kappa|)$. The same bound is also valid for

$$\|\partial(\xi_t^{(r)}, \kappa)\sigma(t, x_t(q), q)\|.$$

Owing to (2.2.3) we conclude that the integrand in (2.2.8) is bounded by an integrable function independent of $r$. That it tends to zero as $r \to 0$ for any $t$ follows from the mean value theorem, the continuity of the first-order derivatives of $\sigma$, and (2.2.7). This and the dominated convergence theorem leads to (2.2.8).

Similarly, for any $R, T,$ and $\omega$

$$(2.2.9) \qquad \int_0^{T \wedge \gamma^{(r)}(R)} |b^{(r)}(t, \xi_t^{(r)}) - \partial(\xi_t^{(r)}, \kappa)b(t, x_t(q), q)| \, dt \to 0$$

as $r \to 0$. Now Theorem 2.1.10 implies that for any $T \in [0, \infty)$ and sequence $r_k \to 0$ we have

$$P - \lim_{k \to \infty} \sup_{t \le T} |\xi_t^{(r_k)} - \xi_t| = 0.$$

This certainly means that $\xi_t = x_{t(\kappa)}(q)$ and proves the first assertion of the present theorem for $m = 1$.

Notice that equation (2.2.4) combined with (2.2.1) composes a system which can be considered as a stochastic equation relative to the couple $x_t, \xi_t$. By the previous result the solution of this system-equation is $t$-locally uniformly differentiable if $m \ge 2$ and as above one can write an equation for its derivative. This equation turns out to be system (2.2.4)-(2.2.5). One can obviously keep going to higher values of $m$, so that the first two assertions of the theorem are proved. One also can avoid considering the equations for higher-order derivatives following the idea introduced in [12] of defining the class of $m$-times differentiable functions as the class of functions whose first derivatives are $m - 1$ times differentiable and using the induction on $m$.

To prove the last assertion of the theorem, we first observe that $x_t(q)$ is $t$-locally uniformly continuous in probability just because it is $t$-locally uniformly differentiable. Next let $q_k \to q_0 \in Q$. Observe that

$$\xi_t^k := x_{t(\kappa)}(q_n) \quad \text{and} \quad \xi_t := x_{t(\kappa)}(q_0)$$

satisfy

$$d\xi_t^k = \sigma^n(t, \xi_t^k)\,dw_t + b^n(t, \xi_t^k)\,dt, \quad d\xi_t = \sigma(t, \xi_t)\,dw_t + b(t, \xi_t)\,dt$$

where

$$\sigma^k(t, \xi) = \partial(\xi, \kappa)\sigma(t, x_t(q_k), q_k), \quad b^k(t, \xi) = \partial(\xi, \kappa)b(t, x_t(q_k), q_k),$$
$$\sigma(t, \xi) = \partial(\xi, \kappa)\sigma(t, x_t(q_0), q_0), \quad b(t, \xi) = \partial(\xi, \kappa)b(t, x_t(q_0), q_0).$$

Define

$$\gamma^k(R) = \inf\{t \geq 0 : |\xi_t^k| \geq R\},$$

and notice that by virtue of $t$-local uniform continuity of $x_t(q)$ in probability (and condition (2.2.3)) for any $T, R$ we have

$$\int_0^{T \wedge \gamma^k(R)} \left[\|\sigma^k - \sigma\|^2 + |b^k - b|\right](t, \xi_t^k)\,dt$$

$$\leq NR \sum_{i,j} \int_0^{T \wedge \gamma^k(R)} \left[\|\sigma_{x^i}(t, x_t(q_k), q_k) - \sigma_{x^i}(t, x_t(q_0), q_0)\|^2\right.$$
$$+ \|\sigma_{q^j}(t, x_t(q_k), q_k) - \sigma_{q^j}(t, x_t(q_0), q_0)\|^2$$
$$+ |b_{x^i}(t, x_t(q_k), q_k) - b_{x^i}(t, x_t(q_0), q_0)|$$
$$\left. + |b_{q^j}(t, x_t(q_k), q_k) - b_{q^j}(t, x_t(q_0), q_0)|\right]\,dt \xrightarrow{P} 0,$$

where $N$ depends only on $d$ and $\kappa$. Also $\xi_0^k \xrightarrow{P} \xi_0$ by assumption.

Hence, by Theorem 2.1.10 we get that $\xi_t^k \to \xi_t$ $t$-locally uniformly in probability, which means that the first $t$-locally uniform derivatives of $x_t(q)$ are $t$-locally uniformly continuous in probability. By applying this conclusion to the couple $(x_t(q), x_{t(\kappa)}(q))$, which satisfies (2.2.1)-(2.2.4), we see that its first $t$-locally uniform derivatives are $t$-locally uniformly continuous in probability if $m \geq 2$. This means that the second $t$-locally uniform derivatives of $x_t(q)$ are $t$-locally uniformly continuous in probability if $m \geq 2$. We apply again this result to the couple $(x_t(q), x_{t(\kappa)}(q))$ and in an obvious way by induction we get that the $m$th $t$-locally uniform derivatives of $x_t(q)$ are $t$-locally uniformly continuous in probability. The theorem is proved. $\square$

EXAMPLE 2.2.4. Let $d = d_1 = 1$ and consider the one-dimensional equation

$$dx_t = dw_t + \frac{1}{x_t}\,dt$$

in $D = \mathbb{R} \setminus \{0\}$. Its solution never leaves $D$, because it is the radial part of a two-dimensional Brownian motion. By Theorem 2.2.3, this solution is infinitely $t$-locally uniformly differentiable in probability with respect to the initial data.

## 2.3.  Random time change in stochastic integrals, I

Let $(w_t, \mathcal{F}_t)$ be a one-dimensional Wiener process. By $\mathcal{P} = \mathcal{P}(\mathcal{F}_\cdot)$ we denote the $\sigma$-algebra of subsets of $\Omega \times (0, \infty)$ generated by the stochastic intervals

$$(0, \tau]] := \{(\omega, t) : 0 < t \le \tau(\omega)\}$$

when $\tau$ runs through the set of all $\mathcal{F}_t$-stopping times. The $\sigma$-algebra $\mathcal{P}$ is called the $\sigma$-algebra of predictable sets. The completion of $\mathcal{P}$ with respect to the measure $P(d\omega) \times \ell$, where $\ell$ is Lebesgue measure on $(0, \infty)$, is denoted by $\mathcal{P}^\mu$. Obviously, $\mathcal{P}^\mu$ is a subset of the completion of $\mathcal{F} \times \mathfrak{B}((0, \infty))$ with respect to $P(d\omega) \times \ell$, where $\mathfrak{B}((0, \infty))$ is the $\sigma$-algebra of Borel subsets of $(0, \infty)$. Therefore, if $A \in \mathcal{P}^\mu$ then by Fubini's theorem for almost any $\omega$ we have

(2.3.1)                         $\{t : (\omega, t) \in A\} \in \mathfrak{B}^\ell((0, \infty))$,

where $\mathfrak{B}^\ell((0, \infty))$ is the completion of $\mathfrak{B}(0, \infty)$ with respect to $\ell$.

It turns out quite inconvenient to work with the sets for which (2.3.1) holds only for almost all $\omega$. Therefore we call a set $A \in \mathcal{P}^\mu$ *pseudopredictable* if (2.3.1) holds for *any* $\omega$ and by $\mathcal{P}_\mu$ we denote the $\sigma$-algebra of all pseudo-predictable sets. Needless to say that the $\mathcal{P}_\mu$-measurable functions are called pseudopredictable.

For $p \in [1, \infty)$ we denote by $H_p$ the space of all pseudopredictable functions $g$ with values in $[-\infty, \infty]$ such that

$$\int_0^T |g(\omega, t)|^p \, dt < \infty$$

for any $\omega \in \Omega$ and $T \in [0, \infty)$. The classes of functions $H_p$ obviously depend on the filtration of $\sigma$-algebras $\{\mathcal{F}_t\}$. We will write $H_p(\mathcal{F}_\cdot)$ for $H_p$, in order to make clear which filtration of $\sigma$-algebras underlies the construction of $H_p$, because we will be varying it in this section. Similarly, we write $\mathcal{P}(\mathcal{F}_\cdot)$ for $\mathcal{P}$, etc.

Let us fix a function $\beta \in H_1(\mathcal{F}_\cdot)$ and assume that for all $\omega, t$

(2.3.2)                 $\beta(\omega, t) > 0,$            $\int_0^\infty \beta(\omega, r) \, dr = \infty.$

Set

(2.3.3)             $\varphi(s) = \int_0^s \beta(t) \, dt,$        $\psi(s) = \inf \{t \ge 0 : \varphi(t) \ge s\}.$

It follows from (2.3.2) that $\varphi(s)$ is continuous and strictly increasing from 0 to $\infty$, $\psi(s)$ is finite and varies continuously from 0 to $\infty$ for $s \in [0, \infty)$. Besides, we know that $\psi(s)$ are stopping times as exit times of a continuous $\mathcal{F}_t$-adapted process. Hence, the $\sigma$-algebras $\mathcal{F}_{\psi(s)}$ are defined and form a filtration of $\sigma$-algebras. These $\sigma$-algebras are complete, because $\mathcal{F}_{\psi(s)} \supset \mathcal{F}_0$. We also note that

$$\varphi(\psi(s)) = \psi(\varphi(s)) = s, \qquad \varphi(s) = \inf\{t \geq 0 : \psi(t) \geq s\},$$

and by the last formula, $\varphi(s)$ is a stopping time for $\{\mathcal{F}_{\psi(t)}\}$ for any $s \in [0, \infty)$.

REMARK 2.3.1. We want to show that not everything is so easy as it may look. Let

$$G_t = \mathcal{F}_{\psi(t)}$$

and consider $G_{\varphi(t)}$. Naturally, we would expect that

$$G_{\varphi(t)} = \mathcal{F}_{\psi(\varphi(t))} = \mathcal{F}_t.$$

However, this need not be true, and the reader is asked to verify this, starting with the case in which

$$\Omega = [1, 2]^2, \quad P = \ell, \quad \mathcal{F}_t = \mathcal{F} = \mathfrak{B}^\ell(\Omega) \quad \text{for} \quad t > 1,$$

$\mathcal{F}_t$ is the completion of

$$\mathfrak{B}([1, 2]) \otimes \{\varnothing, [1, 2]\}$$

for $t \leq 1$, $\beta(\omega_1, \omega_2, t) = \omega_1$. It turns out that in this case $G_{\varphi(1)} = \mathcal{F}_2$.

As a rule, the general theory of stochastic analysis deals with a filtration of $\sigma$-algebras $\mathcal{F}_t$, which is right-continuous in $t$, passing if necessary from $\mathcal{F}_t$ to

$$\mathcal{F}_{t+} := \bigcap_{s > t} \mathcal{F}_s.$$

It is possible to show (see Lemma 2.4.12) that the assumption $\mathcal{F}_t = \mathcal{F}_{t+}$ for all $t \geq 0$ guarantees equality $G_{\varphi(t)} = \mathcal{F}_t$, but we will need neither this assumption nor this equality.

LEMMA 2.3.2.
(a) *If $\tau$ is a stopping time for $\{\mathcal{F}_t\}$, then $\varphi(\tau)$ is a stopping time for $\{\mathcal{F}_{\psi(t)}\}$.*
(b) *A function $f(t)$ is $\{\mathcal{F}_t\}$-predictable if and only if $f(\psi(t))$ is $\{\mathcal{F}_{\psi(t)}\}$-predictable.*
(c) *A continuous process $\xi_t$ is $\mathcal{F}_t$-adapted if and only if $\xi_{\psi(t)}$ is $\mathcal{F}_{\psi(t)}$-adapted.*
(d) *$\beta^{1/p} f(\cdot) \in H_p(\mathcal{F})$ if and only if $f(\psi(\cdot)) \in H_p(\mathcal{F}_{\psi(\cdot)})$, and if $\beta f \in H_1(\mathcal{F})$ then for all $\omega \in \Omega$ and $s \in [0, \infty)$*

$$(2.3.4) \qquad \int_0^{\varphi(s)} f(\psi(t)) \, dt = \int_0^s f(t)\beta(t) \, dt, \quad \int_0^s f(\psi(t)) \, dt = \int_0^{\psi(s)} f(t)\beta(t) \, dt;$$

*in particular (for $f = \beta^{-1}$),*

$$\psi(s) = \int_0^s \beta^{-1}(\psi(t)) \, dt.$$

PROOF. Assertion (a) follows from the fact that

$$\{\varphi(\tau) \leq t\} = \{\tau \leq \psi(t)\} \in \mathcal{F}_{\psi(t)} \cap \mathcal{F}_\tau \subset \mathcal{F}_{\psi(t)}.$$

By virtue of (a), if $f(t) = I_{t \leq \tau}$, where $\tau$ is a stopping time, then

$$f(\psi(t)) = I_{t \leq \varphi(\tau)}$$

is $\{\mathcal{F}_{\psi(t)}\}$-predictable. Hence, by the lemma on $\pi$- and $\lambda$-systems, we easily conclude that for any $\{\mathcal{F}_t\}$-predictable set $A$ and $f = I_A$, the function $f(\psi(t))$ is $\{\mathcal{F}_{\psi(t)}\}$-predictable. Standard approximation of measurable functions by simple functions now proves the "only if" part of assertion (b). Conversely, if $\tau$ is a stopping time relative to $\{\mathcal{F}_{\psi(t)}\}$, then $\{\tau < s\} \in \mathcal{F}_{\psi(s)}$ and

$$\{\tau < s \leq \varphi(t)\} = \{\tau < s, \psi(s) \leq t\} \in \mathcal{F}_t,$$

(2.3.5) $$\{\tau < \varphi(t)\} = \bigcup_{s \in \rho} \{\tau < s \leq \varphi(t)\} \in \mathcal{F}_t, \quad \{\varphi(t) \leq \tau\} \in \mathcal{F}_t,$$

where $\rho$ is the set of all rational points in $[0, \infty)$. Hence, putting $g(t) := I_{t \leq \tau}$, we see that $g(\varphi(t))$ is $\mathcal{F}_t$-adapted, and, since it is also left-continuous, it follows that then $g(\varphi(t))$ is $\{\mathcal{F}_t\}$-predictable. As above, the transition to an arbitrary $\{\mathcal{F}_{\psi(t)}\}$-predictable function $g$ is carried out by standard means and this proves (b).

Assertion (c) is a direct consequence of (b), since for continuous, and for that matter even for left continuous, processes to be $\{\mathcal{F}_t\}$-predictable is the same as to be $\mathcal{F}_t$-adapted.

To prove (d) we observe, that equalities (2.3.4) hold for any fixed $\omega$ and $s$ and any *Borel* function $f(t) \geq 0$ (which is independent of $\omega$). Indeed, the second equality follows from the first one by the substitution $s \to \psi(s)$. The first equality is true for $f(t) = I_{(0,u)}(t)$ since

$$f(\psi(t)) = I_{\psi(t) < u} = I_{t < \varphi(u)}$$

so that

$$\int_0^{\varphi(s)} f(\psi(t)) \, dt = \varphi(u) \wedge \varphi(s),$$

$$\int_0^s f(t)\beta(t) \, dt = \int_0^{s \wedge u} \beta(t) \, dt = \varphi(s \wedge u) = \varphi(u) \wedge \varphi(s).$$

A standard argument immediately extends the first equality in (2.3.4) it to all Borel functions $f \geq 0$.

In addition, if $f$ is still independent of $\omega$, nonnegative and $\mathfrak{B}^\ell((0, \infty))$-measurable, then there exist Borel functions $h, g \geq 0$ such that

$$g(t) = 0 \quad \text{and} \quad g(t)\beta(t) = 0 \quad \text{(a.e.)}, \quad |f - h| \leq g.$$

By (2.3.4),

$$g(\psi(t)) = 0, \quad f(\psi(t)) = h(\psi(t)), \quad f(t)\beta(t) = h(t)\beta(t)$$

(a.e.), and as equalities (2.3.4) are true for $h$, they are also true for $f$. We have also proved that $f(\psi(\cdot))$ is $\mathfrak{B}^\ell((0, \infty))$-measurable for any $\omega$. This is naturally true not only for nonnegative but for any $\mathfrak{B}^\ell((0, \infty))$-measurable function $f$.

On the other hand, if $f(\psi(\cdot))$ is $\mathfrak{B}^\ell((0, \infty))$-measurable for some $\omega$, then there exist Borel functions $g, h$ such that

$$g(t) = 0 \quad \text{(a.e.)}, \quad |f(\psi(t)) - h(t)| \le g(t).$$

Then $h(\varphi), g(\varphi)$ are Borel measurable,

$$g(\varphi(t))\beta(t) = 0 \quad \text{(a.e.)}$$

owing to (2.3.4),

$$g(\varphi(t)) = 0 \quad \text{(a.e.)}$$

due to (2.3.2), implying that

$$f(t) = h(\varphi(t)) \quad \text{(a.e.)}$$

and $f(t)$ is $\mathfrak{B}^\ell((0, \infty))$-measurable for the same $\omega$.

Our definitions now show that, to prove (d), it is sufficient to establish that $f$ is $\mathcal{P}^\mu(\mathcal{F}.)$-measurable if and only if $f(\psi)$ is $\mathcal{P}^\mu(\mathcal{F}_{\psi(\cdot)})$-measurable. In order to do this, it is enough to repeat the argument of the two previous paragraphs with slight changes: take $g, h$ to be not Borel functions, but predictable, use assertion (b) instead of the measurability of a superposition of Borel functions, and use Fubini's theorem along with (2.3.4). The lemma is proved. $\qquad\square$

THEOREM 2.3.3.
(a) *The process*

$$(2.3.6) \qquad \tilde{w}_s := \int_0^{\psi(s)} \beta^{1/2}(t)\, dw_t, \quad s \in [0, \infty),$$

*is a Wiener process with respect to* $\{\mathcal{F}_{\psi(s)}\}$.
(b) *If* $\beta^{\frac{1}{2}} f \in H_2(\mathcal{F}.)$, *then* $f(\psi) \in H_2(\mathcal{F}_{\psi(\cdot)})$ *and*

$$(2.3.7) \qquad \int_0^{\psi(s)} f(t)\beta^{1/2}(t)\, dw_t = \int_0^s f(\psi(t))\, d\tilde{w}_t$$

*(a.s.) for all* $s \in [0, \infty)$ *at once. In particular, for* $f = \beta^{-1/2}$,

$$(2.3.8) \qquad w_{\psi(s)} = \int_0^s \beta^{-1/2}(\psi(t))\, d\tilde{w}_t.$$

PROOF. (a) We prove that $\tilde{w}_t$ is an $\{\mathcal{F}_{\psi(t)}\}$-martingale with $\{\mathcal{F}_{\psi(t)}\}$-quadratic variation equal to $t$. Then since $\tilde{w}_t$ is obviously continuous in $t$ (a.s.), Lévy's theorem would yield the result.

Observe that by Lemma 2 the process $\tilde{w}_s$ is $\mathcal{F}_{\psi(s)}$-adapted. If $\tau$ is a bounded stopping time for $\{\mathcal{F}_{\psi(s)}\}$, then although, generally speaking, $\psi(\tau)$ is not a stopping time for $\{\mathcal{F}_t\}$, nevertheless

$$\{\psi(\tau) < t\} \in \mathcal{F}_t$$

by (2.3.5), and since for any $\varepsilon > 0$, $\tau + \varepsilon$ is also a stopping time for $\mathcal{F}_{\psi(s)}$, it holds that

$$\{\psi(\tau + \varepsilon) \leq t\} = \bigcap_{n=1}^{\infty} \left\{ \psi\left(\tau + \varepsilon - \frac{1}{n}\right) < t \right\} \in \mathcal{F}_t,$$

which shows that $\psi(\tau + \varepsilon)$ *is* a stopping time for $\{\mathcal{F}_t\}$. The following formulas are also valid:

(2.3.9)
$$\int_0^{\infty} \left| I_{t \leq \psi(\tau)} \beta^{1/2}(t) \right|^2 dt = \varphi(\psi(\tau)) = \tau,$$
$$\int_0^{\infty} \left| I_{t \leq \psi(\tau+\varepsilon)} - I_{t \leq \psi(\tau)} \right|^2 \beta(t)\, dt = \int_{\psi(\tau)}^{\psi(\tau+\varepsilon)} \beta(t)\, dt = \varepsilon,$$

from which we see that

$$I_{[0,\psi(\tau+\varepsilon)[\!]} \beta^{1/2}, \quad I_{[0,\psi(\tau)[\!]} \beta^{1/2} \in \mathcal{L}_2(\mathcal{P}(\mathcal{F}.), \mu),$$
$$\tilde{w}_\tau = \lim_{n \to \infty} \tilde{w}_{\tau + \frac{1}{n}} = \lim_{n \to \infty} \int_0^{\infty} I_{t \leq \psi(\tau + \frac{1}{n})} \beta^{1/2}(t)\, dw_t$$

and finally conclude that

(2.3.10)
$$\tilde{w}_\tau = \int_0^{\infty} I_{t \leq \psi(\tau)} \beta^{1/2}(t)\, dw_t \qquad \text{(a.s.)}.$$

We derived (2.3.10) for any bounded $\{\mathcal{F}_{\psi(t)}\}$-stopping time $\tau$. For such $\tau$ (see (2.3.9))

$$E \int_0^{\infty} I_{t \leq \psi(\tau)} \beta(t)\, dt = E\tau < \infty.$$

By the Wald identities we infer from (2.3.10) that

$$E\tilde{w}_\tau = E \int_0^{\infty} I_{t \leq \psi(\tau)} \beta^{1/2}(t)\, dw_t = 0, \qquad E\tilde{w}_\tau^2 = E\tau.$$

Here the first equation along with the $\mathcal{F}_{\psi(t)}$-adaptness of $\tilde{w}_t$ means exactly that $\tilde{w}_t$ is a martingale with respect to $\{\mathcal{F}_{\psi(t)}\}$. The second equation means that

$\tilde{w}_t - t$ is a martingale with respect to $\{\mathcal{F}_{\psi(t)}\}$, so that $\langle \tilde{w} \rangle_s = s$. This finishes the proof of (a) as it was explained above.

(b) Denote the left-hand side of (2.3.7) by $\xi_s$. Arguments analogous to the previous ones show that $\xi_s$ is an $\mathcal{F}_{\psi(s)}$-adapted process, and if $f$ is *bounded* and $\tau$ a bounded stopping time for $\{\mathcal{F}_{\psi(s)}\}$, then

$$I_{(0, \psi(\tau)]} f \beta^{1/2} \in \mathcal{L}_2 \big( \mathcal{P}(\mathcal{F}_{\cdot}), \mu \big)$$

$$\xi_\tau = \int_0^\infty I_{t \leq \psi(\tau)} f(t) \beta^{1/2}(t) \, dw_t \qquad \text{(a.s.)}.$$

Together with the Wald identities, this shows that $\xi_s + \lambda \tilde{w}_s$ (for any constant $\lambda$) and the pair $(\xi_s, \tilde{w}_s)$ are local martingales with respect to $\{\mathcal{F}_{\psi(s)}\}$, and by (2.3.4)

$$\langle \xi \rangle_s = \int_0^{\psi(s)} f^2(t) \beta(t) \, dt = \int_0^s f^2(\psi(t)) \, dt \,,$$

$$\langle \xi, \tilde{w} \rangle_s = \int_0^{\psi(s)} f(t) \beta(t) \, dt = \int_0^s f(\psi(t)) \, dt \,.$$

By Definition III.10.4 of [12], these two equalities imply (2.3.7). Finally, we can prove (2.3.7) for arbitrary $f$ by putting $f_n = (-n) \vee (f \wedge n)$, letting $n \to \infty$ and using (2.3.4) and Theorem III.6.6 of [12].     □

REMARK 2.3.4.  Theorem 2.3.3 remains valid if $w$ is a multidimensional Wiener process and $f$ a function with values in the set of matrices of appropriate dimension. We will not prove this elementary generalization, for the sake of brevity and because we address similar issues in the next section.

## 2.4.  Random time change in stochastic integrals, II

Here we come back to the general situation when $(w_t, \mathcal{F}_t)$ is a $d_1$-dimensional Wiener process. In this section we generalize some results of Section 2.3 in the case in which $\beta$ is defined by

$$\beta = \zeta^{-1} \quad (0^{-1} := \infty) \,,$$

where the process $\zeta = \zeta(t)$ is $[0, \infty)$-valued and *predictable*. With this $\beta$ we borrow other objects from Section 2.3. Also for $\varepsilon > 0$ we define

$$\zeta_\varepsilon = \zeta + \varepsilon, \quad \beta_\varepsilon = \zeta_\varepsilon^{-1}, \quad \varphi_\varepsilon(s) = \int_0^s \beta_\varepsilon(t) \, dt \,.$$

Finally, set

$$\mathcal{F}_t^+ = \mathcal{F}_{t+} \,.$$

As is well known, $\varphi(s)$ is $\mathcal{F}_s$-adapted. However, generally $\varphi(s)$ need not be continuous in $s$. On the other hand, by the monotone convergence theorem

$$\varphi_\varepsilon(s) \uparrow \varphi(s) \quad \text{as} \quad \varepsilon \downarrow 0.$$

It follows that $\varphi(s)$ is lower semicontinuous and, since it is increasing, $\varphi(s)$ is left continuous on $(0, \infty)$. For any $\omega$, as for general lower semicontinuous functions, the sets

$$\{u \geq 0 : \varphi(u) \leq s\}$$

are closed. Moreover, for any $T \in [0, \infty]$, if $\varphi(T) < \infty$, then $\varphi(s)$ is continuous and strictly increasing on $[0, T]$. One of easy consequences of this fact is that $\psi(s)$ is a *continuous* function on $[0, \sup \varphi)$ and if $\sup \varphi < \infty$, then $\psi(s) \to \infty$ as $s \uparrow \sup \varphi$, so that always $\psi$ is *left continuous* on $(0, \infty)$.

REMARK 2.4.1. For any $s \in [0, \infty)$, $u \in [0, \infty]$, $\omega \in \Omega$ we have

(2.4.1)                         $u \leq \psi(s) \Longleftrightarrow \varphi(u) \leq s$.

Indeed, if $\varphi(u) > s$, then $u$ belongs to the open set $\{t : \varphi(t) > s\}$, so that there is a point $t < u$ for which $\varphi(t) > s$. In that case $\psi(s) \leq t < u$. This shows that the left inequality in (2.4.1) implies the right one.

To show the converse observe that if $\varphi(u) \leq s$ and $u = 0$, then the first inequality in (2.4.1) is fulfilled obviously. However, if $\varphi(u) \leq s$ and $u > 0$, then $\varphi$ is strictly increasing on $[0, u]$ and $\varphi(t) < s$ for all $t < u$ showing that $\psi(s) \geq u$.

REMARK 2.4.2. Set

$$\gamma = \psi(\infty) := \inf\{t \geq 0 : \varphi(t) = \infty\}, \quad \kappa = \varphi(\gamma).$$

Then it is not necessarily true that $\kappa = \infty$. However,

(2.4.2)                         $\gamma = \psi(\kappa) = \lim_{n \to \infty} \psi(n)$,

so that if $t > \kappa$, then $\psi(t) = \gamma$.

Indeed, if $\kappa = \infty$, then on $[0, \gamma]$ the function $\varphi$ is continuous, strictly increasing, and takes an infinite value at the right end point. In that case (2.4.2) is obvious.

In the case that $\kappa < \infty$ and $\gamma = \infty$ we have

$$\varphi(s) < \varphi(\infty) = \kappa$$

and

$$\psi(t) = \infty = \gamma,$$

whenever $t \geq \kappa$, so that (2.4.2) holds again.

In the remaining case in which $\kappa < \infty$ and $\gamma < \infty$ we have

$$\varphi(\gamma + \varepsilon) = \infty$$

for any $\varepsilon > 0$. Therefore, $\psi \leq \gamma + \varepsilon$ and $\psi \leq \gamma$. Hence

$$\psi(\kappa) \leq \lim_{n \to \infty} \psi(n) \leq \gamma.$$

On the other hand $\psi(\kappa) = \gamma$ since $\varphi(\gamma) = \kappa$ and $\varphi(s) < \kappa$ if $s < \gamma$ due to the fact that on $[0, \gamma]$ the function $\varphi$ is continuous, strictly increasing and finite.

REMARK 2.4.3. If $t < \kappa$, that is $t < \varphi(\gamma)$, then by using (2.4.1) we get $\psi(t) < \psi(\kappa)$ and $\psi(t) < \infty$. It follows from the definition of $\psi$ that on $[0, \psi(t)]$ the function $\varphi$ is continuous, strictly increasing and takes the value $t$ at the right end point. Hence,

$$t < \kappa \implies \psi(t) < \psi(\kappa), \quad \varphi(\psi(t)) = t.$$

The equality $\varphi(\psi(t)) = t$ further implies that

$$\zeta(s) > 0 \quad \text{on} \quad [0, \psi(t)] \quad \text{(a.e.)}.$$

REMARK 2.4.4. If $t < \kappa$ then by Remark 2.4.3

$$t \wedge \kappa = \varphi(\psi(t)).$$

This equality also holds if $t \geq \kappa$ since then $\psi(t) = \psi(\kappa) = \gamma$ by Remark 2.4.2 and $\varphi(\gamma) = \kappa$.

REMARK 2.4.5. The definition of $\psi$ implies that $\psi(\varphi(t)) \leq t$ for any $t \in [0, \infty]$. This proves the implication $\implies$ in

$$s \wedge \kappa \leq \varphi(t) \iff \psi(s \wedge \kappa) \leq t,$$

which we assert to hold for all $s, t \in [0, \infty)$. The opposite implication follows from Remark 2.4.4.

REMARK 2.4.6. It follows from (2.4.1) and the fact that $\varphi(u)$ is $\mathcal{F}_u$-adapted that for any $s, u \in [0, \infty)$

$$\{\omega : \psi(s) \geq u\} = \{\omega : \varphi(u) \leq s\} \in \mathcal{F}_u,$$
$$\{\omega : \psi(s) > u\} = \bigcap_{n \geq 1} \{\omega : \psi(s) \geq u + 1/n\} \in \mathcal{F}_u^+.$$

Hence $\psi(s)$ is an $\{\mathcal{F}_t^+\}$-stopping time for any $s \in [0, \infty)$, by (2.4.2) the same is true for $s = \infty$, and therefore the $\sigma$-algebras

$$\hat{\mathcal{F}}_s := \mathcal{F}_{\psi(s)}^+$$

are well defined for all $s \in [0, \infty]$.

Furthermore, $\kappa$ is an $\{\hat{\mathcal{F}}_s\}$-stopping time since by (2.4.1) and by the fact that both $\psi(t)$ and $\gamma$ are $\{\mathcal{F}_t^+\}$-stopping times, for any $t \in [0, \infty)$,

$$\{\omega : \varphi(\gamma) > t\} = \{\omega : \gamma > \psi(t)\} \in \mathcal{F}_{\psi(t)}^+ = \hat{\mathcal{F}}_t .$$

DEFINITION 2.4.7. Let $\tau$ be an $\{\mathcal{F}_t\}$-stopping time and $\eta_t$ be an $\mathbb{R}^{d_1}$-dimensional continuous process defined on $[0, \tau] \cap [0, \infty)$. We say that $\eta_t$ is a local $\{\mathcal{F}_t\}$-martingale on $[0, \tau)$ if there is a sequence of stopping times $\tau_n \leq \tau$ such that $\tau_n \to \tau$ and for any $n$ the process $\eta_{t \wedge \tau_n}$ is an $\{\mathcal{F}_t\}$-martingale. If $\eta_t$ is a local $\{\mathcal{F}_t\}$-martingale on $[0, \tau)$, one introduces its quadratic variation $\langle \eta \rangle_t$, $t \leq \tau$, as usually.

If $\eta_t$ is a local $\{\mathcal{F}_t\}$-martingale on $[0, \tau)$, we say that $\eta_t$ is an $\mathbb{R}^{d_1}$-dimensional Wiener process on $[0, \tau)$ relative to $\{\mathcal{F}_t\}$ if $\langle \eta, \eta \rangle_t = It$, $t \leq \tau$, where $I$ is the identity matrix of appropriate dimensions.

The following result will be used few times in the future.

LEMMA 2.4.8. *Let $\tau$ be a stopping time such that $P(\tau < \infty) = 1$ and let $\eta_t$ be a continuous random process defined on $[0, \tau]$ such that the family $\{\eta_{t \wedge \tau}, t \geq 0\}$ is uniformly integrable. Then the following statements are equivalent:*

(i) $\eta_t$ *is a local martingale on $[0, \tau)$,*
(ii) $\eta_{t \wedge \tau}$ *is a martingale on $[0, \infty)$,*
(iii) $\eta_{t \wedge \tau}$ *is $\mathcal{F}_t$-adapted and for any stopping time $\pi \leq \tau$*

$$(2.4.3) \qquad\qquad\qquad E\eta_\pi = E\eta_0 .$$

PROOF. The implications (i) $\Longrightarrow$ (ii) $\Longrightarrow$ (iii) are straightforward. That (iii) $\Longrightarrow$ (i) is also well known. By the way, in [12] property (iii) with bounded $\kappa$ in (2.4.3) is taken as the definition of martingale. This allows one to avoid using conditional expectations. The lemma is proved.                                    □

THEOREM 2.4.9. *Let $\sigma(t)$ and $b(t)$ be $\mathcal{P}(\mathcal{F}.)$-measurable processes with values in the set of $d \times d_1$-matrices and $\mathbb{R}^d$, respectively. Assume that*

$$(2.4.4) \qquad\qquad \int_0^T (\|\sigma(t)\|^2 + |b(t)|) \, dt < \infty$$

*for any $\omega \in \Omega$ and $T \in [0, \infty)$ and introduce*

$$\xi(t) = \int_0^t \sigma(s) \, dw_s + \int_0^t b(s) \, ds .$$

*Then*

(i) *there exists a continuous $\{\hat{\mathcal{F}}_t\}$-martingale $\hat{w}_t$ on $[0, \infty)$ such that*

$$(2.4.5) \qquad\qquad \hat{w}_t = \hat{w}_{t \wedge \kappa}, \quad \forall t \in [0, \infty)$$

(a.s.), $\hat{w}_t$ is a $d_1$-dimensional Wiener process on $[0, \kappa)$ relative to $\{\hat{\mathcal{F}}_t\}$ and for any bounded $\{\hat{\mathcal{F}}_t\}$-stopping time $\tau$

(2.4.6) $$\hat{w}_\tau = \int_0^\infty \zeta^{-1/2}(s) I_{s \leq \psi(\tau)} \, dw_s \quad (a.s.) ;$$

(ii) with probability one for all $n = 1, 2, \ldots$ and $s \in [0, \infty)$

(2.4.7)
$$\xi(\psi(s) \wedge n) = \int_0^s (\zeta^{1/2}\sigma)(\psi(r)) I_{r \leq \varphi(n)} \, d\hat{w}_r$$
$$+ \int_0^s (\zeta b)(\psi(r)) I_{r \leq \varphi(n)} \, d(r \wedge \kappa) ;$$

(iii) if

(2.4.8) $$\int_0^{\psi(s)} \|\sigma(t)\|^2 \, dt < \infty \quad (a.s.) \quad \forall s \in [0, \infty) ,$$

then (a.s.) for all $s \in [0, \infty)$

(2.4.9) $$\xi(\psi(s)) = \int_0^s (\zeta^{1/2}\sigma)(\psi(r)) \, d\hat{w}_r + \int_0^s (\zeta b)(\psi(r)) \, dr .$$

REMARK 2.4.10. Condition (2.4.8) is *equivalent* to the fact that

$$\int_0^{\psi(s)} \sigma(t) \, dw_t$$

is well defined for all $s \in [0, \infty)$ even on the sets of $\omega$ where $\psi(s) = \infty$ when it is understood as

$$\lim_{n \to \infty} \int_0^n \sigma(t) \, dw_t ,$$

which exists and is finite (a.s.) (see, for instance, [12]).

Therefore, without condition (2.4.8) the left-hand side of (2.4.9) makes no sense. The same is true for the stochastic integral on the right. Indeed, in view of assertion (i) and formula (2.4.12) below the corresponding quadratic variation is $d_1$ times

$$\int_0^s \|(\zeta\sigma)(\psi(r))\|^2 \, d(r \wedge \kappa) = \int_0^{s \wedge \kappa} \|(\zeta\sigma)(\psi(r))\|^2 \, dr = \int_0^{\psi(s)} \|\sigma(t)\|^2 \, dt$$

and the latter is infinite if (2.4.8) is violated.

REMARK 2.4.11. In view of (2.4.4) condition (2.4.8) is satisfied if $\psi(s) < \infty$ (a.s.) for any $s \in [0, \infty)$. The latter is, obviously, equivalent to

$$\varphi(\infty) = \infty \quad (a.s.) .$$

To prove the theorem and, in particular, show that the stochastic integrals in (2.4.6) and (2.4.7) make sense, we need the following lemma somewhat similar to Lemma 2.3.2.

LEMMA 2.4.12.

(a) *If $\tau$ is a stopping time for $\{\mathcal{F}_t\}$, then $\varphi(\tau)$ is a stopping time for $\{\hat{\mathcal{F}}_t\}$,*

$$(2.4.10) \qquad \hat{\mathcal{F}}_{\varphi(\tau)} \subset \mathcal{F}_\tau^+ \quad \text{provided that} \quad P(\varphi(\tau) < \infty) = 1, \quad \hat{\mathcal{F}}_{\varphi(\tau)} \supset \mathcal{F}_\tau^+,$$

*and if $\theta$ is an $\{\hat{\mathcal{F}}_t\}$-stopping time, then $\psi(\theta)$ is an $\{\mathcal{F}_t^+\}$-stopping time.*

(b) *If a function $f(t)$ is $\{\mathcal{F}_t\}$-predictable, then $f(\psi(t))$ is $\{\hat{\mathcal{F}}_t\}$-predictable.*

(c) *If a continuous process $\xi_t$ is $\mathcal{F}_t$-adapted, then $\xi_{\psi(t)}$ is $\hat{\mathcal{F}}_t$-adapted.*

(d) *If a function $f(t)$ is $\{\mathcal{F}_t\}$-predictable and nonnegative, then for all $\omega \in \Omega$*

$$(2.4.11) \qquad \int_0^{\varphi(s)} f(\psi(t))\, dt = \int_0^s f(t)\beta(t)\, dt \quad \forall s \in [0, \gamma],$$

$$(2.4.12) \qquad \int_0^{s \wedge \kappa} (\zeta f)(\psi(t))\, dt = \int_0^{\psi(s)} f(t)\, dt \quad \forall s \in [0, \infty].$$

*In both (2.4.11) and (2.4.12) expressions $0 \cdot \infty$ whenever they occur are set to be 0.*

PROOF. (a) As above by (2.4.1) for any $s \geq 0$

$$\{\omega : \varphi(\tau) \leq s\} = \{\omega : \tau \leq \psi(s)\}$$

and the latter set is in $\hat{\mathcal{F}}_s$, since both $\psi(s)$ and $\tau$ are $\{\mathcal{F}_t^+\}$-stopping times. Hence $\varphi(\tau)$ is a stopping time for $\{\hat{\mathcal{F}}_t\}$.

Next we prove (2.4.10). It is worth saying that, although (2.4.10) answers few very natural questions, it will never be used in these notes.

Take $A \in \hat{\mathcal{F}}_{\varphi(\tau)}$ and assume that $\varphi(\tau) < \infty$ (a.s.). Then by definition for any finite $s, u \geq 0$, $t > 0$

$$A \cap \{\omega : \varphi(\tau) \leq s\} \in \hat{\mathcal{F}}_s = \mathcal{F}_{\psi(s)}^+,$$

$$A \cap \{\omega : \varphi(\tau) \leq s, \psi(s) \leq u\} \in \mathcal{F}_u^+.$$

Observe that

$$(2.4.13) \qquad \{\omega : \varphi(\tau) \leq s, \psi(s) \leq u\} = \{\omega : \tau \leq \psi(s) \leq u\}.$$

Therefore,

$$A \cap \{\omega : \tau \leq \psi(s) \leq u\} \in \mathcal{F}_u^+, \qquad A \cap \{\omega : \tau \leq \psi(s) < t\}$$

$$= \bigcup_{n \geq 1} A \cap \{\omega : \tau \leq \psi(s) \leq t - 1/n\} \in \bigcup_{n \geq 1} \mathcal{F}_{t-1/n}^+ \subset \mathcal{F}_t.$$

It follows that

$$(2.4.14) \qquad A \cap \bigcup_r \{\omega : \tau \leq \psi(r) < t\} \in \mathcal{F}_t,$$

where $r$ runs through the set of rational numbers on $[0, \infty)$.

It turns out that

$$\bigcup_r \{\omega : \tau \le \psi(r) < t\} = \{\omega : \tau < t\} \quad \text{(a.s.)}.$$

Indeed, the left-hand side is obviously a subset of the set on the right. On the other hand, if on an $\omega$ we have $\tau < t$ and $\varphi(\tau) < \infty$ then either $\varphi(\tau+) < \infty$ or $\varphi(\tau+) = \infty$. In the second case

$$\gamma = \tau, \quad \psi(r) = \tau \quad \forall r > \varphi(\tau)$$

and this $\omega$ belongs to the left-hand side. The same is true if $\varphi(\tau+) < \infty$, because then there is a $T > \tau$ such that $\varphi(T) < \infty$ and since $\varphi$ is continuous and strictly increasing on $[0, T]$ there is a rational number $r$ such that $\psi(r)$ is as close to $\tau$ from the right as we wish. In particular, again we can find $r$ so that

$$\tau < \psi(r) < t.$$

Now (2.4.14) shows that

$$A \cap \{\omega : \tau < t\} \in \mathcal{F}_t, \quad A \cap \{\omega : \tau \le u\} = \bigcap_{n \ge 1} A \cap \{\omega : \tau < u + 1/n\} \in \mathcal{F}_u^+$$

and we just have proved that

$$\hat{\mathcal{F}}_{\varphi(\tau)} \subset \mathcal{F}_\tau^+ \quad \text{if} \quad P(\varphi(\tau) < \infty) = 1.$$

To prove the remaining inclusion in (2.4.10), take an $A \in \mathcal{F}_\tau^+$. Then for $s \in [0, \infty)$

$$A \cap \{\omega : \tau \le s\} \in \mathcal{F}_s^+$$

and for any $\{\mathcal{F}_t^+\}$-stopping time $\pi$

(2.4.15)
$$A \cap \{\omega : \tau \le \pi\} \in \mathcal{F}_\pi^+$$

since

$$A \cap \{\omega : \tau \le \pi\} = \bigcap_{n \ge 1} A \cap \{\omega : \tau < \pi + 1/n\},$$

$$A \cap \{\omega : \tau < \pi + 1/n, \pi \le s\}$$
$$= \bigcup_{r \le s + 1/n} A \cap \{\omega : \tau \le r\}\{\omega : r < \pi + 1/n, \pi \le s\} \in \mathcal{F}_{s+1/n},$$

and

$$A \cap \{\omega : \tau \le \pi, \pi \le s\} \in \bigcap_{n \ge 1} \mathcal{F}_{s+1/n} = \mathcal{F}_s^+.$$

Iterating (2.4.15) yields

$$A \cap \{\omega : \tau \leq \pi \leq u\} \in \mathcal{F}_u^+, \quad u \in [0, \infty).$$

Take here $\pi = \psi(s)$, which even is an $\{\mathcal{F}_t\}$-stopping time, and use (2.4.13) to obtain

$$A \cap \{\omega : \varphi(\tau) \leq s, \psi(s) \leq u\} \in \mathcal{F}_u^+.$$

After that the arbitrariness of $u$ means that

$$A \cap \{\omega : \varphi(\tau) \leq s\} \in \mathcal{F}_{\psi(s)}^+ = \hat{\mathcal{F}}_s,$$

and the arbitrariness of $s$ that $A \in \hat{\mathcal{F}}_{\varphi(\tau)}$. This finishes the proof of (2.4.10). To prove the last assertion in (a) observe that for $t > 0$

$$\{\omega : \psi(\theta) < t\} = \{\omega : \theta < \varphi(t)\} = \bigcup_{r \geq 0} \{\omega : \theta \leq r < \varphi(t)\},$$

where $r$ runs through the set of rational numbers on $[0, \infty)$. Here by definition

$$\{\omega : \theta \leq r\} \in \mathcal{F}_{\psi(r)}^+, \quad \{\omega : \theta \leq r < \varphi(t)\} = \{\omega : \theta \leq r, \psi(r) < t\}$$
$$= \bigcup_{n \geq 1} \{\omega : \theta \leq r, \psi(r) \leq t - 1/n\} \in \bigcup_{n \geq 1} \mathcal{F}_{t-1/n}^+ \subset \mathcal{F}_t.$$

Hence, for $t > 0$ and $s \geq 0$

$$\{\omega : \psi(\theta) < t\} \in \mathcal{F}_t, \quad \{\omega : \psi(\theta) \leq s\}$$
$$= \bigcap_{n \geq 1} \{\omega : \psi(\theta) < s + 1/n\} \in \bigcap_{n \geq 1} \mathcal{F}_{s+1/n} = \mathcal{F}_s^+,$$

meaning that $\psi(\theta)$ is an $\{\mathcal{F}_t^+\}$-stopping time.

(b) It suffices to consider $f(t) = I_{t \leq \tau}$, where $\tau$ is an $\{\mathcal{F}_t\}$-stopping time. In that case

$$f(\psi(t)) = I_{\psi(t) \leq \tau}$$

is $\hat{\mathcal{F}}_t$-measurable as in (a) and also left continuous as a superposition of a left-continuous function $f$ with a left-continuous increasing function $\psi$. Hence $f(\psi(t))$ is $\{\hat{\mathcal{F}}_t\}$-predictable indeed.

Assertion (c) is a general property of stopped adapted processes.

We prove (d) for each particular $\omega$ and Borel $f$. For $s < \gamma$ when $\varphi(s) < \infty$ (and $\beta < \infty$ (a.e.) on $[0, s]$) equation (2.4.11) is proved in the same way as in Lemma 2.3.2. For $s = \gamma$ it holds by the left continuity of both parts of the equation provided that $\gamma > 0$. However, if $\gamma = 0$, then (2.4.11) holds automatically.

By substituting $\psi(s)$ in place of $s$ in (2.4.11) and using Remark 2.4.4 and the fact that $\psi \le \gamma$ we obtain

$$\int_0^{s \wedge \kappa} f(\psi(t)) \, dt = \int_0^{\psi(s)} f(t)\beta(t) \, dt \, .$$

Now, for $f$ taking only finite values (2.4.12) is proved by the same substitution as in Lemma 2.3.2. For general $f$ it suffices to use the monotone convergence theorem. The lemma is proved. $\qquad\square$

PROOF OF THEOREM 2.4.9. Obviously it suffices to consider two cases:

(a) $\sigma \equiv 0$,
(b) $b \equiv 0$.

In (a) due to (2.4.1) if $s < \varphi(\infty)$, then $\psi(s) < \infty$, and (2.4.12) and (2.4.4) imply that

$$\infty > \int_0^n |b(t)| \, dt \ge \int_0^{\psi(s)\wedge n} |b(t)| \, dt = \int_0^{\psi(s)} |b(t)|I_{t \le n} \, dt$$

$$= \int_0^{s \wedge \kappa} |\zeta b|(\psi(t))I_{\psi(t)\le n} \, dt = \int_0^s |\zeta b|(\psi(t))I_{\psi(t\wedge\kappa)\le n} \, d(t \wedge \kappa) \, .$$

Next, use Remark 2.4.5 to get that

$$\infty > \int_0^{\psi(s)\wedge n} |b(t)| \, dt = \int_0^s |\zeta b|(\psi(t))I_{t\wedge\kappa\le\varphi(n)} \, d(t \wedge \kappa)$$

$$= \int_0^s |\zeta b|(\psi(t))I_{t\le\varphi(n)} \, d(t \wedge \kappa) \, .$$

It follows that

$$\int_0^s (\zeta b)(\psi(t))I_{t\le\varphi(n)} \, d(t \wedge \kappa) = \int_0^{\psi(s)\wedge n} b(t) \, dt \, ,$$

which settles case (a) in what concerns assertion (ii). Assertion (iii) is proved by setting $n \to \infty$ in the above argument and noticing that $\kappa \le \varphi(\infty)$.

In the rest of the proof we assume that $b \equiv 0$. For $n = 1, 2, \ldots$ and $t \in [0, \infty)$ define

(2.4.16) $\quad m_t^n = \int_0^t \zeta^{-1/2}(s)I_{s\le\psi(n)} \, dw_s, \quad \hat{w}_t^n = m_{\psi(t)}^n, \quad \hat{w}_t = \lim_{n\to\infty} \hat{w}_t^n \, .$

A justification of the definition is in order. By (2.4.12)

(2.4.17) $\quad \int_0^{t\wedge\psi(n)} \zeta^{-1}(s) \, ds \le \int_0^{\psi(n)} \zeta^{-1}(s) \, ds$

$$= \int_0^{n\wedge\kappa} (\zeta\zeta^{-1})(\psi(t)) \, dt = \int_0^{n\wedge\kappa} I_{\zeta\ne0}(\psi(t)) \, dt \le n \, .$$

We see that the quadratic variation corresponding to the stochastic integral in (2.4.16) is bounded, hence the stochastic integral exists and defines a uniformly integrable $\{\mathcal{F}_t\}$-martingale on $[0, \infty]$. In particular $m_\infty^n$ makes perfect sense and so does $m_{\psi(t)}^n$ for any $t \in [0, \infty]$.

Also for any $t \in [0, \infty]$

$$(2.4.18) \qquad\qquad m_{\psi(t)}^n = \lim_{s \to t} m_{\psi(s)}^n .$$

Indeed, since $\psi$ is left continuous, it suffices to concentrate on $s \downarrow t$. However, if $\psi(t) = \gamma$, then by Remark 2.4.2 we have $\psi(t) = \sup \psi$ and $\psi(s) = \sup \psi = \gamma$, so that

$$m_{\psi(s)}^n = m_{\psi(t)}^n .$$

Furthermore, if $\psi(t) < \gamma$, then $t < \infty$ and by (2.4.1) we have $t < \varphi(\gamma) = \kappa$. Then for $s > t$ close enough to $t$ it holds that $s < \kappa$, which by Remark 2.4.3 implies that $\varphi$ is continuous and strictly increasing on $[0, \psi(s)]$ and therefore $\psi$ is continuous on $[0, s]$. In that case (2.4.18) follows by the continuity of $m_r^n$.

Next observe that $\hat{w}_t^n$ is $\{\hat{\mathcal{F}}_t\}$-adapted as it follows from Lemma 2.4.12 (c) and the fact that $m_t^n$ is $\{\mathcal{F}_t\}$-adapted. Furthermore, if $\pi$ is a bounded stopping time for $\{\hat{\mathcal{F}}_t\}$, then as above

$$\int_0^{\psi(\pi)} \zeta^{-1}(s)\, ds \le \pi$$

implying that $\langle m^n \rangle_{\psi(\pi)}$ is bounded and

$$E w_\pi^n = E m_{\psi(\pi)}^n = 0 .$$

Hence $\hat{w}_t^n$ are continuous $\{\hat{\mathcal{F}}_t\}$-martingales.

To find their quadratic variations we need the following formula: for any $t \in [0, \infty)$ and $\omega$

$$(2.4.19) \qquad\qquad \int_0^{\psi(t)} \zeta^{-1}(s)\, ds = t \wedge \kappa .$$

To prove (2.4.19) notice that by Lemma 2.4.12

$$(2.4.20) \qquad\qquad \int_0^{\psi(t)} \zeta^{-1}(s)\, ds = \int_0^{t \wedge \kappa} (\zeta \zeta^{-1})(\psi(r))\, dr .$$

By Remark 2.4.3 for any $\omega$ and $t < \kappa$

$$\zeta(s) > 0 \quad \text{on} \quad [0, \psi(t)) \quad (\text{a.e.}).$$

By letting $t \uparrow \kappa$ and using Remark 2.4.2 and the left continuity of $\psi$ we conclude

$$(2.4.21) \qquad\qquad \zeta(s) > 0 \quad \text{on} \quad [0, \gamma)$$

(a.e.) for any $\omega$.

Now for fixed $\omega$ we redefine if needed $\zeta$ on a Borel set of measure zero in such a way that for the new $\zeta$ equation (2.4.21) would hold on the whole of $[0, \gamma]$ rather than only almost everywhere. This change will not affect $\varphi(s)$ as long as $s \leq \gamma$ due to (2.4.21). Consequently, it will not affect $\psi(r)$ for $r \leq \varphi(\gamma) = \kappa$. But now the last integrand in (2.4.20) identically equals 1 for $r \leq \kappa$. This proves (2.4.19).

In light of (2.4.19) for $k \geq n$

$$E\langle \hat{w}^n - \hat{w}^k \rangle_t = d_1 E \int_0^{\psi(t)} \zeta^{-1}(s) |I_{s \leq \psi(n)} - I_{s \leq \psi(k)}| \, ds$$

$$= d_1 E \int_{\psi(n) \wedge \psi(t)}^{\psi(k) \wedge \psi(t)} \zeta^{-1}(s) \, ds \leq d_1 (k \wedge t - n \wedge t) \to 0$$

as $n \to \infty$. It follows that as $n \to \infty$ the martingales $\hat{w}_t^n$ converge $t$-locally uniformly in probability to a continuous martingale, so that the limit in (2.4.16) is well defined and $\hat{w}_t$ is a continuous martingale on $[0, \infty)$. Its quadratic variation is the limit of the quadratic variations of $\hat{w}_t^n$ and formula (2.4.19) shows that

$$\langle \hat{w}, \hat{w} \rangle_t = I(t \wedge \kappa),$$

where $I$ is the $d_1 \times d_1$ identity matrix. Upon taking any bounded $\{\hat{\mathcal{F}}_t\}$-stopping time $\tau$ and noting that

$$E \left| \hat{w}_\tau - \int_0^\infty \zeta^{-1/2}(s) I_{s \leq \psi(\tau)} \, dw_s \right|^2$$

$$= \lim_{n \to \infty} E \left| \int_0^\infty \zeta^{-1/2}(s) I_{s \leq \psi(\tau) \wedge \psi(n)} \, dw_s - \int_0^\infty \zeta^{-1/2}(s) I_{s \leq \psi(\tau)} \, dw_s \right|^2$$

$$= d_1 \lim_{n \to \infty} E \int_{\psi(\tau) \wedge \psi(n)}^{\psi(\tau)} \zeta^{-1}(s) \, ds \leq d_1 \lim_{n \to \infty} E(\tau - \tau \wedge n) = 0,$$

we obtain (2.4.6). Finally, writing

$$E |\hat{w}_t - \hat{w}_{t \wedge \kappa}|^2 = d_1 E(t \wedge \kappa - (t \wedge \kappa) \wedge \kappa) = 0,$$

we get (2.4.5) and finish proving assertion (i) of the theorem.

To prove assertion (ii) we first prove that if instead of (2.4.4) for any $s \in [0, \infty)$ we have

(2.4.22) $$\int_0^{\psi(s)} \zeta^{-1}(t) \|\sigma(t)\|^2 \, dt < \infty$$

then for any $s \in [0, \infty)$ with probability one

(2.4.23) $$\int_0^\infty \zeta^{-1/2}(t) \sigma(t) I_{t \leq \psi(s)} \, dw_t = \int_0^s \sigma(\psi(r)) \, d\hat{w}_r.$$

It suffices to do that when all the processes are one-dimensional and $\sigma$ is real valued.

First notice that both parts of (2.4.23) are well defined. For the left-hand side this follows directly from (2.4.22) and for the right-hand side we need only use Remark 2.4.4, by which

$$s \wedge \kappa = \varphi(\psi(s)),$$

and $\psi(s) \leq \gamma$, so that by (2.4.11)

$$\int_0^s |\sigma(\psi(r))|^2 d(r \wedge \kappa) = \int_0^{s \wedge \kappa} |\sigma(\psi(r))|^2 dr$$
$$= \int_0^{\varphi(\psi(s))} |\sigma(\psi(r))|^2 dr = \int_0^{\psi(s)} \zeta^{-1}(t) \|\sigma(t)\|^2 dt < \infty.$$

This computation and also the possibility to appropriately approximate any predictable $\sigma$ with linear combinations of indicators of stochastic intervals shows that it suffices to prove (2.4.23) for $\sigma = I_{(0,\tau]}$, where $\tau$ is a bounded $\{\mathcal{F}_t\}$-stopping time. Just in case, observe that then for any $u \in [0, \infty)$

$$\infty > u \geq \varphi(\tau) \wedge u \wedge \kappa = \varphi(\tau \wedge \psi(u)) = \int_0^{\psi(u)} \zeta^{-1}(t) I_{t \leq \tau} dt,$$

so that (2.4.22) is automatically satisfied.

For $\sigma = I_{(0,\tau]}$ owing to (2.4.5) and Remark 2.4.5 the right-hand side of (2.4.23) turns out to be (a.s.)

(2.4.24)
$$\int_0^s I_{\psi(r) \leq \tau} d\hat{w}_r = \int_0^s I_{\psi(r) \leq \tau} d\hat{w}_{r \wedge \kappa} = \int_0^s I_{\psi(r \wedge \kappa) \leq \tau} d\hat{w}_r$$
$$= \int_0^s I_{r \wedge \kappa \leq \varphi(\tau)} d\hat{w}_r = \int_0^s I_{r \leq \varphi(\tau)} d\hat{w}_r = \hat{w}_{\varphi(\tau) \wedge s}.$$

By using (2.4.6) and (2.4.1) we further continue the above equalities

$$\hat{w}_{\varphi(\tau) \wedge s} = \int_0^\infty \zeta^{-1/2}(r) I_{r \leq \psi(\varphi(\tau) \wedge s)} dw_r$$
$$= \int_0^\infty \zeta^{-1/2}(r) I_{\varphi(r) \leq \varphi(\tau) \wedge s} dw_r = \int_0^\infty \zeta^{-1/2}(r) I_{r \leq \tau \wedge \psi(s)} dw_r,$$

where in the last equality we also used that $\varphi(r) \leq s$ implies that $r \leq \psi(s)$ and $\varphi(r) < \infty$, which combined with $\varphi(r) \leq \varphi(\tau)$ yields that $r \leq \tau$. This proves (2.4.23).

Finally, take $\sigma$ satisfying (2.4.4) and $n = 1, 2, \ldots$ and substitute $\zeta^{1/2} \sigma I_{[0,n]}$ into (2.4.23) in place of $\sigma$. Since for any $s \in [0, \infty)$ we have (use again (2.4.21))

$$\int_0^{\psi(s)} \zeta^{-1}(t) \|\zeta^{1/2}(t) \sigma(t)\|^2 I_{t \leq n} dt = \int_0^{\psi(s) \wedge n} \|\sigma(t)\|^2 dt < \infty$$

equation (2.4.23) yields

$$(2.4.25) \qquad \int_0^\infty \sigma(t) I_{t \le \psi(s) \wedge n} \, dw_t = \int_0^s \zeta^{1/2}(\psi(r)) \sigma(\psi(r)) I_{\psi(r) \le n} \, d\hat{w}_r$$

for any $s \in [0, \infty)$ with probability one. Here the left-hand side equals

$$\xi(\psi(s) \wedge n)$$

(a.s.) and on the right we can replace $\psi(r) \le n$ with $r \le \varphi(n)$ as in (2.4.24). It follows that for any $s \in [0, \infty)$

$$\xi(\psi(s) \wedge n) = \int_0^s \zeta^{1/2}(\psi(r)) \sigma(\psi(r)) I_{r \le \varphi(n)} \, d\hat{w}_r$$

almost surely. Since both parts are left-continuous functions of $s$, almost surely the equality holds for all $s$ at once. This proves assertion (ii).

As is explained in Remark 2.4.10, under the condition in assertion (iii) both parts of equation (2.4.25) make sense if we replace $n$ with $\infty$. Then standard results from the theory of stochastic integration imply that after setting $n \to \infty$ we get (2.4.25) with the said substitution. By the standard results we mean that the convergence of stochastic integrals is equivalent to the convergence of their quadratic variations and we have, for instance,

$$\int_0^s \zeta(\psi(r)) \| \sigma(\psi(r)) \|^2 \, d(r \wedge \kappa) = \int_0^{s \wedge \kappa} (\zeta \| \sigma \|^2)(\psi(r)) \, dr$$

$$= \int_0^{\psi(s)} \| \sigma(t) \|^2 \, dt < \infty, \quad \int_0^s (\zeta \| \sigma \|^2)(\psi(r)) I_{r > \varphi(n)} \, d(r \wedge \kappa) \to 0$$

as $n \to \infty$ by the dominated convergence theorem. After that we finish proving (iii) in the same way as (ii).

This brings the proof of the theorem to an end. □

## 2.5. Random time change in Itô stochastic equations

Assume that, for any $x \in \mathbb{R}^d$, we are given a $d \times d_1$-dimensional matrix $\sigma(x)$ and a $d$-vector $b(x)$. We assume that $\sigma$ and $b$ are Borel and the equation

$$(2.5.1) \qquad x_t = \int_0^t \sigma(x_s) \, dw_s + \int_0^t b(x_s) \, ds, \quad t \ge 0,$$

has a solution, which means that there exists a continuous $\mathcal{F}_t$-adapted $\mathbb{R}^d$-valued process $x_t$ such that

$$(2.5.2) \qquad \int_0^T (\| \sigma(x_t) \|^2 + |b(x_t)|) \, dt < \infty$$

(a.s.) for any $T \in [0, \infty)$ and (2.5.1) is satisfied.

It is worth noting that we consider $\sigma$ and $b$ depending only on $x$ only for simplicity of notation. The reader will have no trouble in adjusting our proofs for the most general case, when notation become somewhat cumbersome.

Next, we take a *bounded continuous* function $\zeta$ on $\mathbb{R}^d$ and introduce

$$D = \{x : \zeta(x) > 0\},$$

$$\varphi(t) = \int_0^t \zeta^{-1}(x_s)\,ds, \quad \psi(s) = \inf\{t \geq 0 : \varphi(t) \geq s\}$$

with usual agreement that

$$0^{-1} = \infty, \quad \inf\{\emptyset\} = \infty.$$

Observe that since $\zeta(x_s)$ is bounded, $\zeta^{-1}(x_s)$ is bounded away from zero and $\varphi(\infty) = \infty$. Therefore, by Remark 2.4.11 and Theorem 2.4.9 the process

$$y_s := x_{\psi(s)}$$

is well defined and satisfies the equation

$$(2.5.3) \qquad y_t = \int_0^t (\zeta^{1/2}\sigma)(y_s)\,d\hat{w}_s + \int_0^t (\zeta b)(y_s)\,d(s \wedge \kappa), \quad t \geq 0,$$

where $\hat{w}_t$ is a continuous $\{\hat{\mathcal{F}}_t\}$-martingale with

$$\langle \hat{w}, \hat{w} \rangle_t = I(t \wedge \kappa), \quad \hat{\mathcal{F}}_t = \mathcal{F}^+_{\psi(t)}, \quad \mathcal{F}^+_t = \mathcal{F}_{t+}, \quad \kappa = \varphi(\gamma), \quad \gamma = \psi(\infty).$$

REMARK 2.5.1. It holds that (a.s.)

$$(2.5.4) \qquad\qquad y_t = y_{t \wedge \kappa}, \quad \forall t \in [0, \infty),$$

and

$$0 \in \bar{D} \Longrightarrow y_t \in \bar{D} \quad \forall t \in [0, \infty).$$

Indeed, since $\hat{w}_t = \hat{w}_{t \wedge \kappa}$ we have

$$\int_0^t (\zeta^{1/2}\sigma)(y_s)\,d\hat{w}_s = \int_0^t (\zeta^{1/2}\sigma)(y_s)\,d\hat{w}_{s \wedge \kappa} = \int_0^{t \wedge \kappa} (\zeta^{1/2}\sigma)(y_s)\,d\hat{w}_s.$$

Similar equation for the second term on the right in (2.5.3) is obvious and this proves (2.5.4).

If for a $t \in [0, \infty)$ we assume that $y_t \notin \bar{D}$, then

$$t > 0, \quad x_{\psi(t)} \notin \bar{D}, \quad \psi(t) > 0$$

and there is an $\varepsilon \in (0, \psi(t)]$ such that $x_s \notin \bar{D}$ for $s \in [\psi(t) - \varepsilon, \psi(t)]$. Then $\varphi(s) = \infty$ for $s \in [\psi(t) - \varepsilon, \psi(t)]$, implying that

$$\gamma \leq \psi(t) - \varepsilon < \infty, \quad \gamma < \infty.$$

However, since $\psi(t) \leq \psi(\infty) = \gamma$, the inequality $\gamma \leq \psi(t) - \varepsilon$ is impossible.

Set

$$\tau = \inf\{t \geq 0 : x_t \notin D\}.$$

REMARK 2.5.2. If a function $v$ defined in $\bar{D}$ is such that $v(x_t)$ is a local $\{\mathcal{F}_t\}$-martingale on $[0, \tau)$, then $v(y_t)$ is a local $\{\hat{\mathcal{F}}_t\}$-martingale on $[0, \varphi(\tau))$.

To prove that, let $\tau_n \leq \tau$ be a localizing sequence for $v(x_t)$. Without losing generality we may assume that each of $\tau_n$ and $\varphi(\tau_n)$ is bounded. Then $v(x_{t \wedge \tau_n})$ are uniformly integrable and even more than just that: for any $\{\mathcal{F}_t\}$-stopping time $\pi$ and constant $c > 0$

$$E|v(x_{\pi \wedge \tau_n})|I_{|v(x_{\pi \wedge \tau_n})| \geq c} \leq E|v(x_{\tau_n})|I_{\sup_t |v(x_{t \wedge \tau_n})| \geq c},$$

(2.5.5)
$$P(\sup_t |v(x_{t \wedge \tau_n})| \geq c) \leq \frac{1}{c} E|v(x_{\tau_n})|.$$

By Lemma 2.4.8 for any $\{\mathcal{F}_t\}$-stopping time $\pi$ we have

(2.5.6)
$$Ev(x_{\pi \wedge \tau_n}) = v(0).$$

Equation (2.5.6) also holds for any $\{\mathcal{F}_t^+\}$-stopping time $\pi$, because then $\pi + \varepsilon$ is an $\{\mathcal{F}_t\}$-stopping time for any $\varepsilon > 0$, we can substitute $\pi + \varepsilon$ in place of $\pi$ and then pass to the limit by using the uniform integrability and the continuity of $v(x_t)$.

Now take an $\{\hat{\mathcal{F}}_t\}$-stopping time $\hat{\pi}$ and introduce

$$\hat{\tau}_n = \varphi(\tau_n), \quad \pi = \psi(\hat{\pi}).$$

By Lemma 2.4.12 it holds that $\hat{\tau}_n$ are $\{\hat{\mathcal{F}}_t\}$-stopping times and $\pi$ is an $\{\mathcal{F}_t^+\}$-stopping time. Furthermore, $\tau_n = \psi(\hat{\tau}_n)$ since $\varphi(\tau_n) < \infty$ and on $[0, \tau_n]$ the function $\varphi$ is continuous and strictly increasing.

Finally, (2.5.6) yields

$$Ev(y_{\hat{\pi} \wedge \hat{\tau}_n}) = v(0)$$

and it only remains to use Lemma 2.4.8, an obvious relation: $\hat{\tau}_n \to \varphi(\tau)$, and the fact that the uniform integrability of $v(y_{t \wedge \hat{\tau}_n})$ follows from (2.5.5).

THEOREM 2.5.3. *Assume that $\zeta$ is twice continuously differentiable, its first and second derivatives are bounded, and $\sigma$ and $b$ are bounded on $\bar{D}$. Also let $0 \in D$. Then*

(i)
$$y_t \in D \quad \forall t \in [0, \infty), \quad \gamma = \tau, \quad \kappa = \infty \quad (a.s.),$$

*so that*

$$y_t = \int_0^t (\zeta^{1/2}\sigma)(y_s) \, d\hat{w}_s + \int_0^t (\zeta b)(y_s) \, ds, \quad t \geq 0,$$

*where $\hat{w}_t$ is a $d_1$-dimensional Wiener process relative to $\{\hat{\mathcal{F}}_t\}$;*

(ii) *if a function $v$ defined in $\bar{D}$ is such that $v(x_t)$ is a local $\{\mathcal{F}_t\}$-martingale on $[0, \tau)$, then $v(y_t)$ is a local $\{\hat{\mathcal{F}}_t\}$-martingale on $[0, \infty)$.*

PROOF. (i) By Itô's formula

$$(2.5.7) \qquad d\zeta(y_t) = \zeta(y_t)A(y_t)\, d(t \wedge \kappa) + \zeta(y_t)B(y_t)\, d\hat{w}_t ,$$

where

$$A = a^{ij}\zeta_{x^i x^j} + b^i\zeta_{x^i}, \quad a = (1/2)\sigma\sigma^*,$$
$$B \in \mathbb{R}^{d_1}, \quad B^k = \zeta_{x^i}\zeta^{1/2}\sigma^{ik}/\zeta \quad (0/0 := 0).$$

By assumption $A$ is a bounded function in $\bar{D}$ and by Remark 2.5.1 we obtain that $A(y_t)$ is a bounded process.

Furthermore, it is a simple and well-known fact that if $u$ is twice continuously differentiable nonnegative function of one variable, then

$$|u'| \le 2u^{1/2}\sup|u''|^{1/2}$$

pointwise. It follows that $B$ is also a bounded function in $\bar{D}$ and $B(y_t)$ is a bounded process.

In such situation $\zeta(y_t)$ as a solution of (2.5.7) is written as

$$\zeta(y_t) = \zeta(0)\exp\left(\int_0^t B(y_s)\, d\hat{w}_s\right.$$
$$\left. -(1/2)\int_0^t |B(y_s)|^2\, d(s \wedge \kappa) + \int_0^t A(y_s)\, d(s \wedge \kappa)\right),$$

which shows that $\zeta(y_t) > 0$ and $y_t \in D$ for any $t \in [0, \infty)$ and $x_r \in D$ for $r \in [0, \psi(\infty))$ (a.s.). Hence

$$(2.5.8) \qquad \psi(\infty) = \gamma \le \tau \quad \text{(a.s.)}.$$

On the other hand, on $[0, \tau)$ the function $\varphi$ is finite. Therefore,

$$\tau \le \gamma = \inf\{t \ge 0 : \varphi(t) = \infty\}.$$

Thus, $\tau = \gamma$ and

$$\varphi(\tau) = \varphi(\gamma) = \kappa$$

(a.s.). To finish proving assertion (i) it only remains to notice that if $\varphi(\tau) = \varphi(\gamma) < \infty$, then

$$\psi(t) = \gamma = \tau$$

for all $t \ge \varphi(\tau)$ and

$$y_t = x_{\psi(t)} = x_\tau \in \partial D,$$

which may only happen with probability zero by the above.

Assertion (ii) is just an adaptation of Remark 2.5.2 to our particular case in which $\varphi(\tau) = \kappa = \infty$. The theorem is proved. $\qquad\square$

## 3. Quasiderivatives

In this chapter we consider the following Itô equation

$$(3.0.1) \qquad x_t = x + \int_0^t \sigma^k(x_s)\, dw_s^k + \int_0^t b(x_s)\, ds \,,$$

on a given complete probability space $(\Omega, \mathcal{F}, P)$, where $x$ is a given point in $\mathbb{R}^d$, $\sigma^k(y)$, $b(y)$ are (nonrandom) functions on $\mathbb{R}^d$ with values in $\mathbb{R}^d$, defined for $k = 1, \ldots, d_1$ with $d_1$ possibly different from $d$, $w_t = (w_t^1, \ldots, w_t^{d_1})$ is a $d_1$-dimensional Wiener process with respect to a given increasing filtration

$$\{\mathcal{F}_t, t \geq 0\}$$

of $\sigma$-algebras $\mathcal{F}_t \subset \mathcal{F}$, such that $\mathcal{F}_t$ contain all $P$ null sets. By $\sigma$ we denote the $d \times d_1$ matrix composed of the column-vectors $\sigma^k$. We also take a domain

$$D \subset \mathbb{R}^d$$

and suppose that the functions $\sigma$ and $b$ are once continuously differentiable and, for any $x \in D$, equation (3.0.1) has a unique solution defined for all $t \in [0, \infty)$, which we denote by $x_t(x)$. For $x \in D$ define

$$\tau(x) = \inf\{t \geq 0 : x_t(x) \notin D\} \quad (\inf\{\emptyset\} : \infty)\,.$$

### 3.1. The notion of quasiderivative

DEFINITION 3.1.1. We write

$$u \in \mathcal{M} \quad (= \mathcal{M}(D, \sigma, b))$$

if $u$ is a real-valued continuously differentiable function given on $\bar{D}$ such that the process $u(x_t(x))$ is a local $\{\mathcal{F}_t\}$-martingale on $[0, \tau(x))$ for any $x \in D$.

Let $x \in D$, $\xi \in \mathbb{R}^d$, let $\tau$ be a stopping time, $\tau \leq \tau(x)$, $\xi_t$ and $\xi_t^0$ adapted continuous processes defined on

$$[0, \tau] \cap [0, \infty)$$

with values in $\mathbb{R}^d$ and $\mathbb{R}$ respectively and such that $\xi_0 = \xi$. We say that $\xi_t$ is a *quasiderivative of $x_t(y)$ in the direction of $\xi$ at point $x$ on $[0, \tau)$* if for any $u \in \mathcal{M}$ the following process

$$(3.1.1) \qquad u_{(\xi_t)}(x_t(x)) + \xi_t^0 u(x_t(x))$$

is a local martingale on $[0, \tau)$. In this case the process $\xi_t^0$ is called *an adjoint process* for $\xi_t$. If $\tau = \tau(x)$ we simply say that $\xi_t$ is a quasiderivative of $x_t(y)$ in $D$ in the direction of $\xi$ at $x$.

EXAMPLE 3.1.2. Assume that $\tau(x) < \infty$ (a.s.) for any $x \in D$, take a Borel bounded $g$ defined on $\partial D$ and introduce

$$u(x) = E_x g(x_\tau).$$

The strong Markov property of $x_t(x)$ implies that

$$u(x_{t \wedge \tau(x)}(x))$$

is a martingale on $[0, \infty)$ for any $x \in D$. Next, assume that $u$ is $C^1$ in $\bar{D}$. Then $u \in \mathcal{M}$. Recall that the difference of local martingales on $[0, \infty)$ is a local martingale on $[0, \infty)$ and the limit in probability of local martingales is again a local martingale provided that the limit is uniform on each finite time interval. This makes it natural that

$$(3.1.2) \qquad \lim_{\varepsilon \to 0} \frac{u(x_t(x + \varepsilon \xi)) - u(x_t(x))}{\varepsilon}$$

is a local martingale on $[0, \tau(x))$. Furthermore, under our standing assumptions we know from Theorem 2.2.3 that $x_t(x)$ is differentiable in $x$ in the sense of uniform convergence of the corresponding ratios in probability on every finite time interval and if we denote by $\xi_t$ the limit of

$$\varepsilon^{-1}(x_t(x + \varepsilon \xi) - x_t(x)),$$

then $\xi_t$ satisfies the following equation which is obtained by formal differentiating (3.0.1)

$$(3.1.3) \qquad \xi_t = \xi + \int_0^t \sigma_{(\xi_s)}^k(x_s)\, dw_s^k + \int_0^t b_{(\xi_s)}(x_s)\, ds.$$

Hence (3.1.2) is just

$$(3.1.4) \qquad u_{(\xi_t)}(x_t(x)),$$

so that usual derivative with respect to $x$ of $x_t(x)$ is a quasiderivative with zero adjoint process. In particular, this shows that the name "quasiderivative" is natural.

However, the above derivation has a flaw. The point is that

$$u(x_t(x + \varepsilon \xi)) \quad \text{and} \quad u(x_t(x))$$

are local martingales on different time intervals and not on $[0, \infty)$. One could avoid this difficulty by stopping both of them at the same moment $\tau$ defined as the infimum of $\tau(y)$ over a small neighborhood of $x$ and then letting the diameter of the neighborhood tend to zero. Yet new unpleasant difficulties could appear related to the infimum of $\tau(y)$ which function is even not continuous in $y$.

There is a technically more convenient way based on random time change. Take a twice continuously differentiable function $\zeta \geq 0$ with compact support and such that $\zeta = 0$ outside of $D$. Then for each $x \in \mathbb{R}^d$ the equation

$$(3.1.5) \qquad \bar{x}_t = x + \int_0^t (\zeta \sigma^k)(\bar{x}_s)\, dw_s^k + \int_0^t (\zeta^2 b)(\bar{x}_s)\, ds\,,$$

has a unique solution which we denote by $\bar{x}_t(x)$.

LEMMA 3.1.3. *If $x \in D$ and $u$ is a continuous function in $\bar{D}$ such that $u(x_t(x))$ is a local martingale on $[0, \tau(x))$, then $u(\bar{x}_t(x))$ is a local martingale on $[0, \infty)$. Furthermore $\bar{x}_t \in D$ for all $t$ (a.s.).*

PROOF. Both assertions are almost immediate consequences of Theorem 2.5.3, which says that the lemma is true if we replace $\bar{x}_t$ with the solution $y_t$ of

$$y_t = x + \int_0^t (\zeta \sigma)(y_s)\, d\hat{w}_s + \int_0^t (\zeta^2 b)(y_s)\, ds\,,$$

with certain Wiener process $\hat{w}_s$ and additionally specify the filtration with respect to which $u(y_t)$ is a local martingale. However, $u(y_t)$ is certainly a local martingale with respect to the filtration generated by $(\hat{w}., y.)$ and this property of $u(y_t)$ is completely described in terms of the finite-dimensional distributions of $(\hat{w}., y.)$. The finite-dimensional distributions of $(\hat{w}., y.)$ and $(w., \bar{x}.(x))$ coincide because the equations defining $\bar{x}.(x)$ and $y.$ have Lipschitz continuous coefficients. In particular, $\bar{x}_t(x)$ never leaves $D$ since the same holds for $y_t$.

It also follows that $u(\bar{x}_t(x))$ is a local martingale with respect to the filtration generated by $(w., \bar{x}.(x))$. Furthermore, $\bar{x}_t(x)$ is measurable with respect to

$$\mathcal{F}_t^w := \sigma(w_s : s \leq t)$$

and hence $u(\bar{x}_t(x))$ is a local martingale with respect to $\{\mathcal{F}_t^w\}$. In addition, $u(\bar{x}_t(x))$ is bounded since $u$ is bounded on the compact support of $\zeta$. Therefore $u(\bar{x}_t(x))$ is a martingale. These properties and the Markov property of $w.$ imply that for $t \geq s$

$$E\{u(\bar{x}_t(x))|\mathcal{F}_s\} = E\{u(\bar{x}_t(x))|\mathcal{F}_s^w\}. = u(\bar{x}_s(x)) \quad \text{(a.s.)}.$$

The lemma is proved.                                                      □

Now we come back to process (3.1.4). Take a bounded domain $D' \subset \bar{D}' \subset D$ and take a nonnegative infinitely differentiable function $\zeta$ such that $\zeta = 1$ on $D'$, $\zeta = 0$ outside $D$. According to Lemma 3.1.3 the process $u(\bar{x}_t(x))$ is a local martingale on $[0, \infty)$. Therefore, our argument about (3.1.4) is valid and

$$(3.1.6) \qquad\qquad u_{(\bar{\xi}_t)}(\bar{x}_t(x))$$

is a local martingale, where $\bar{\xi}_t$ is defined by equation (3.1.3) with $\zeta\sigma$ and $\zeta^2 b$ in place of $\sigma$ and $b$, respectively. Finally, obviously

$$\bar{\xi}_t = \xi_t, \quad \bar{x}_t(x) = x_t(x)$$

before $x_t(x)$ exits from $D'$ and this proves that (3.1.4) is indeed a local martingale on $[0, \tau(x))$.

REMARK 3.1.4. General properties of quasiderivatives can be found in [9].

REMARK 3.1.5. Assume that $u \in \mathcal{M}$, the domain $D$ is smooth, $\tau(x) < \infty$ (a.s.), and $u = g$ on $\partial D$, where $g$ is a given smooth function. If process (3.1.1) is, actually, a uniformly integrable martingale on $[0, \tau(x))$ and $\xi_{\tau(x)}$ is tangent to $\partial D$ at $x_{\tau(x)}(x)$ (a.s.), then we have

$$u_{(\xi)}(x) = E_x[u_{(\xi_\tau)}(x_\tau) + \xi_\tau^0 u(x_\tau)] = E_x[g_{(\xi_\tau)}(x_\tau) + \xi_\tau^0 g(x_\tau)].$$

This shows how we one can use quasiderivatives to get interior estimates of $u_{(\xi)}$ through $|g|_{1,D}$.

A particular case occurs when $\xi = 0$. Then

$$(3.1.7) \qquad \begin{aligned} 0 = u_{(0)}(x) &= E[g_{(\xi_{\tau(x)})}(x_{\tau(x)}) + \xi_{\tau(x)}^0 g(x_{\tau(x)})], \\ Eg_{(\xi_{\tau(x)})}(x_{\tau(x)}) &= -E\xi_{\tau(x)}^0 g(x_{\tau(x)}). \end{aligned}$$

This is a kind of integrating by parts formula. Formulas of this type were widely used by Bismut [2] (1981) in his famous investigation related to Malliavin's calculus and Hörmander's theorem about smoothness of distribution densities of diffusion processes. Note that the distribution of $x_T(x)$ at a given nonrandom time moment $T$ can be viewed as the exit distribution of the diffusion process $(t, x_t(x))$ from the domain $\{(t, x) : t \in (0, T)\}$. At the same time formula (3.1.7) concerns exit distributions from more general domains. Therefore, we can hope that (3.1.7) can be used to prove that under some general conditions exit distributions from smooth domains have smooth densities.

This remark shows a reason for attempting to construct as many quasi-derivatives as possible, and a few of them we present in the following section.

## 3.2. Basic examples of quasiderivatives

The argument in the proof of Lemma 3.1.3 is based on random time change and on the fact that random time change preserves local martingales. There is a simple relation between local martingales when we perform change of measure as in Girsanov's theorem. One can also change the driving Wiener process by using rotations. By combining all these transformations one arrives at the following result.

THEOREM 3.2.1. *Let $r_t$, $\pi_t$, $P_t$ be jointly measurable adapted processes with values in $\mathbb{R}$, $\mathbb{R}^{d_1}$, and in the set of all $d_1 \times d_1$ skew-symmetric matrices, respectively. Assume that*

$$\int_0^T (|r_t|^2 + |\pi_t|^2 + |P_t|^2)\, dt < \infty$$

*for any $T \in [0, \infty)$. For $x \in D$, $\xi \in \mathbb{R}^d$ on the time interval $[0, \infty)$ define the process $\xi_t$ as a solution of the following (linear) equation*

$$
\begin{aligned}
\xi_t = \xi &+ \int_0^t [\sigma_{(\xi_s)} + r_s \sigma + \sigma P_s]\, dw_s \\
&+ \int_0^t [b_{(\xi_s)} + 2r_s b - \sigma \pi_s]\, ds\,,
\end{aligned}
$$

(3.2.1)

*where in $\sigma$, $b$ and their derivatives we dropped the argument $x_s(x)$. Also define*

$$\xi_t^0 = \int_0^t \pi_s \, dw_s \,.$$

(3.2.2)

*Then $\xi_t$ is a quasiderivative of $x_t(y)$ in $D$ in the direction of $\xi$ at $x$ and $\xi_t^0$ is its adjoint process.*

PROOF. By using common cut-off procedure and recalling that the uniform limit in probability of local martingales is a local martingale, we see that while proving that (3.1.1) is a local martingale we may assume that $r_t$, $\pi_t$, and $P_t$ are bounded processes.

Next, as in the argument after the proof of Lemma 3.1.3 we may assume that $\sigma$ and $b$ vanish outside of a compact set in $D$. This reduces the general situation to the case

$$D = \mathbb{R}^d$$

and we may concentrate on bounded $u \in \mathcal{M}$ when local martingales $u(x_t(x))$ actually are martingales.

In this situation take an $\varepsilon$ so small that $1 + 2\varepsilon r_t \geq \varepsilon$, and define the process $y_t(\varepsilon, x)$ as a solution to the equation

$$
\begin{aligned}
dy_t = {}&\sqrt{1 + 2\varepsilon r_t}\, \sigma(y_t) e^{\varepsilon P_t}\, dw_t \\
&+ r_t) b(y_t) - \sqrt{1 + 2\varepsilon r_t}\, \sigma(y_t) e^{\varepsilon P_t} \varepsilon \pi_t]\, dt
\end{aligned}
$$

(3.2.3)

with initial condition $y_t = x + \varepsilon \xi$.

We want to show that for any bounded $u \in \mathcal{M}$ the process

(3.2.4)     $$u(y_t(\varepsilon, x)) \exp\left(\varepsilon \int_0^t \pi_s \, dw_s - \frac{1}{2}\varepsilon^2 \int_0^t |\pi_s|^2 \, ds\right)$$

is a martingale that is (see Lemma 2.4.8)

$$Eu(y_\gamma(\varepsilon, x)) \exp\left(\varepsilon \int_0^\gamma \pi_s \, dw_s - \frac{1}{2}\varepsilon^2 \int_0^\gamma |\pi_s|^2 \, ds\right) = u(x)$$

for any bounded stopping time $\gamma$.

By using the Girsanov theorem we see that it suffices to do this for $\pi_t = 0$ only. Theorem 2.3.3 and an obvious random time change, that does not affect the martingale property, shows that we can put $r_t = 0$ as well. But if $\pi_t = 0$, $r_t = 0$ in (3.2.3), then $y_t(\varepsilon, x)$ has the same finite dimensional distributions as $x_t(x+\varepsilon\xi)$ does, which follows from weak uniqueness of solutions to (3.0.1) and from the fact that

$$\tilde{w}_t := \int_0^t e^{\varepsilon P_s} \, dw_s$$

is a Wiener process for any $\varepsilon$ since $\exp \varepsilon P_s$ is an orthogonal matrix. Thus, process (3.2.4) is indeed a martingale for any small $\varepsilon$, and by differentiation of this process with respect to $\varepsilon$ at point $\varepsilon = 0$ we conclude that process (3.1.1) is at least a local martingale. The theorem is proved.     □

DEFINITION 3.2.2. According to the way equation (3.2.1) is derived we call the quasiderivative time-change related if $\pi_t \equiv 0$, $P_t \equiv 0$ and measure-change related if $r_t \equiv 0$, $P_t \equiv 0$.

REMARK 3.2.3. If we applied a different method of proving (see [9]), we could show that in (3.2.1) instead of $\sigma P$ we can take $\tilde{\sigma}$ for any locally bounded adapted process $\tilde{\sigma}$ taking values in the set of $d \times d_1$ matrices and such that $\sigma\tilde{\sigma}^* \equiv -\tilde{\sigma}\sigma^*$. We believe that with this addition, the theorem describes *all* possible quasiderivatives of solutions to (3.0.1) (with smooth coefficients).

REMARK 3.2.4. Take functions $f(x)$, $c(x)$ continuously differentiable in $D$ and a function $g(x)$ on $\mathbb{R}^d$, and assume that the formula

$$v(x) = E_x \int_0^\tau e^{-\phi_t} f(x_t) \, dt + E_x e^{-\phi_\tau} g(x_\tau),$$

with

$$\phi_t(x) := \int_0^t c(x_s(x)) \, ds,$$

makes sense and defines a function $v$ that is continuously differentiable in $D$. It turns out then that the process

(3.2.5)     $$e^{-\phi_t}[v_{(\xi_t)}(x_t) + \xi_t^0 v(x_t)] + \int_0^t e^{-\phi_s}[f_{(\xi_s)}(x_s) + \xi_s^0 f(x_s)] \, ds$$

is a local martingale on $[0, \tau(x))$, where $x_t = x_t(x)$, $\phi_t = \phi_t(x)$, $\xi_t$ is defined by equation (3.2.1), and

(3.2.6)
$$\xi_t^0 = \int_0^t \pi_s \, dw_s - \int_0^t [c_{(\xi_s)}(x_s) + 2r_s c(x_s)] \, ds \, .$$

Indeed, introduce processes $x_t^{d+1}$, $x_t^{d+2}$ with the help of the following "equations"

$$x_t^{d+1} = x^{d+1} - \int_0^t c(x_s) \, ds, \quad x_t^{d+2} = x^{d+2} + \int_0^t e^{x_s^{d+1}} f(x_s) \, ds \, .$$

Note that the Markov property of $x_t(x)$ implies that the process

$$e^{-\phi_t} v(x_t) + \int_0^t e^{-\phi_s} f(x_s) \, ds$$

is a local martingale on $[0, \tau(x))$, which in terms of the process

$$\tilde{x}_t = (x_t, x_t^{d+1}, x_t^{d+2})$$

means that the process $u(\tilde{x}_t)$ is a local martingale on $[0, \tau(x))$, where

$$u(\tilde{y}) := y^{d+2} + v(y) \exp y^{d+1}$$

for

$$\tilde{y} = (y, y^{d+1}, y^{d+2}), \ y \in D, \ y^{d+1}, y^{d+2} \in \mathbb{R} \, .$$

Now our assertion follows quite formally from Theorem 3.2.1 applied to the process

$$\tilde{x}_t = (x_t, x_t^{d+1}, x_t^{d+2}) \, ,$$

which is a solution of (3.0.1) supplemented with two new equations, to the function $\tilde{u}$, and to the domain

$$\{(y, y^{d+1}, y^{d+2}) : y \in D, \ y^{d+1}, y^{d+2} \in \mathbb{R}\} \, .$$

This remark shows one of reasons why in Theorem 3.2.1 we avoided to impose any growth assumptions on $\sigma$, $b$ when $D$ is unbounded.

## 4. Some examples of applying quasiderivatives

Here we take the objects from the introduction to Chapter 3 a sufficiently smooth bounded function $g$ on $\mathbb{R}^d$, introduce

$$v(x) = E_x g(x_\tau)$$

(see Preface concerning this notation), and *assume* that

(4.0.1)                                      $v \in C^1(\bar{D})$.

Under this conditions we derive few interior estimates of the first-order derivatives of $v$ in terms of either $|g|_{0,D}$ or $|g|_{1,D}$.

Actually, in all the cases considered in this chapter one can add small nondegenerate diffusion to (3.0.1) and mollify $g$ so that one could rely on known results from the theory of PDE which guarantee (4.0.1), then use our estimates and finally pass to the limit making the added diffusion vanish and mollified $g$ converge to the original one. This would prove our results under natural assumptions only. We do not want to do this because the goal of this chapter is not to present some results in their frightening generality but rather show how quasiderivatives work.

## 4.1. Using time-change related quasiderivatives I

EXAMPLE 4.1.1. We come back to the situation of Remark 1.0.2. Take $d = 2, d_1 = 1$, and $D = \{x : |x| < 1\}$. Define the process $x_t(x)$ and the function $v$ by the equations

$$x_t^1 = x^1 + w_t, \quad x_t^2 = x^2, \quad v(x) = E_x g(x_\tau).$$

Note that our process moves only horizontally and one easily obtains that

(4.1.1)        $$v(x) = \frac{x^1}{\sqrt{1 - |x^2|^2}} f\left(\sqrt{1 - |x^2|^2}\right) + h\left(\sqrt{1 - |x^2|^2}\right),$$

where $2f(z) = g(z) - g(-z)$ and $2h(z) = g(z) + g(-z)$. In particular, for $x^1 = 0$

(4.1.2)                        $$v_{x^2}(x) = -\frac{x^2 h'(\sqrt{1 - |x^2|^2})}{\sqrt{1 - |x^2|^2}}$$

and the best estimate we can get in terms of $|g'|_{0,\partial D}$ is

$$|v_{x^2}(x)| \leq \frac{2|x^2|}{\sqrt{1 - |x^2|^2}} |g'|_{0,\partial D}.$$

We see that, if $|g'|_{0,\partial D}$ is under control but $h'(0)$ approaches a nonzero constant, then as $x$ goes to the south pole the derivative of $v(x)$ in $x^2$ at $x = (0, x^2)$ more and more behaves as the inverse to the square root of the distance of $x$ to the boundary of $D$.

We will see that this kind of behavior can be explained probabilistically without using the explicit formula for $v$.

Take $r_t$ to be a constant $r$, $\pi_t \equiv 0$, and $P_t \equiv 0$. Then for $\xi = (0, 1)$ we find that

$$(4.1.3) \qquad v_{x^2}(x) = v_{(\xi)}(x) = E_x v_{(\xi_\tau)}(x_\tau(x)),$$

where the process $\xi_t$ satisfies the equation

$$(4.1.4) \qquad d\xi_t^1 = r\, dw_t, \quad d\xi_t^2 = 0, \quad \xi_0 = \xi.$$

Let us use the freedom in the choice of the parameter $r$ in order to evaluate $v_{x^2}(x)$ for $x = (0, x^2)$. We want to choose $r$ in such a way that the derivative of $v$ *disappears* from the right-hand side of (4.1.3). To do this it suffices to make $\xi_{\tau(x)}$ to be tangent to $\partial D$ at the point $x_{\tau(x)}(x)$. Indeed, in this case we can replace $v$ by $g$ in the right-hand side of (4.1.3).

Thus, we need an $r$ such that

$$\xi_{\tau(x)} = (r w_{\tau(x)}, 1) \perp (x_{\tau(x)}^1, x_{\tau(x)}^2) = (w_{\tau(x)}, x^2),$$

and since

$$w_{\tau(x)}^2 + |x^2|^2 = 1,$$

we (only) can take

$$r = -x^2 (1 - |x^2|^2)^{-1}.$$

By taking this $r$ and observing that, owing to symmetry, $E_x f_{(\xi_\tau)}(x_\tau) = 0$ for $x = (0, x^2)$, we conclude from (4.1.3) that

$$v_{x^2}(x) = E_x h_{(\xi_\tau)}(x_\tau) = E_x h'(|x_\tau^1|)(\operatorname{sgn} x_\tau^1) r w_\tau$$
$$= r E_x h'(\sqrt{1 - |x^2|^2})|x_\tau^1| = r h'(\sqrt{1 - |x^2|^2})\sqrt{1 - |x^2|^2},$$

which is precisely (4.1.2).

Interestingly enough this result can also be obtained on the basis of usual derivatives. The point is that for $x = (0, x^2)$, obviously,

$$v(x) = E_x h(x_\tau^1) =: u(x)$$

and $u(x)$ is constant on the horizontal lines, so that its gradient is vertical and on the boundary is well related to the tangential derivative. Namely,

$$u_{x^2} x^1 - u_{x^1} x^2$$

on the boundary is the tangential derivative equal to $u_{x^2}x^1$. It follows that

$$u_{x^2} = \frac{1}{x^1}[h_{x^2}x^1 - h_{x^1}x^2] = -\frac{x^2}{x^1}h_{x^1}.$$

Now the formula

$$u_{x^2}(x) = E_x u_{x^2}(x_\tau)$$

yields

$$v_{x^2}(x) = u_{x^2}(x) = -x^2 E_x h'(|x_\tau^1|)(\operatorname{sgn} x_\tau^1)(x_\tau^1)^{-1}$$
$$= -x^2 E_x h'(|x_\tau^1|)|x_\tau^1|^{-1} = -x^2 h'(\sqrt{1-|x^2|^2})(1-|x^2|^2)^{-1/2},$$

which is again (4.1.2).

REMARK 4.1.2. Formula (4.1.1) shows that $v$ in $D$ is not smoother than $g$, so that our process does not possess smoothing property and the only source of smoothness for $v$ is on the boundary. One may wonder why then the smoothness of $v$ becomes worse as the point approaches this source.

One of possible explanations is the following. We see that we can turn the quasiderivative in such a way that it becomes tangent to the boundary when and where the phase process hits it. We have to pay for the use of quasiderivatives in the sense that we have to estimate $\xi_{\tau(x)}$, and it is quite natural that the closer the starting point is to the boundary, the fewer time is left until the process hits the boundary, the larger efforts we have to apply to turn the quasiderivative appropriately, and the larger price we have to pay for it. This explains why the derivative $v_{x^2}(0, x^2)$ could blow up (as it does) when $x^2 \to \pm 1$.

Let us use the same idea to treat more general processes (3.0.1) with time dependent coefficients.

THEOREM 4.1.3 ([9]0. *Let* $d \geq 2$,

$$D = \{x : x^d > 0\}, \quad b^d \equiv -1, \quad \sigma^{dk} \equiv 0 \text{ for } i = 1, \ldots, d_1.$$

*Recall that*

$$v(x) = E_x g(x_\tau).$$

*We assert that*

(4.1.5)          $$|v_{(\xi)}(x)| \leq N(|\xi| + (x^d)^{-\frac{1}{2}}|\xi^d|)e^{Nx^d} \sup_{\partial D} |g_x|$$

*in $D$ for any $\xi \in \mathbb{R}^d$ with a constant $N$ depending only on $d$, $d_1$ and bounds for $\sigma$, $b$ and their first derivatives.*

PROOF. Fix $\xi \in \mathbb{R}^d$, a real $r$ and define a quasiderivative $\xi_t$ as a solution to the equation

$$(4.1.6) \qquad d\xi_t = (\sigma_{(\xi_t)} + r\sigma) \, dw_t + (b_{(\xi_t)} + 2rb) \, dt$$

with $\xi_0 = \xi$, where for simplicity we drop the argument $x_t(x)$ in $\sigma$, $b$ and their derivatives. In particular,

$$d\xi_t^d = -2r \, dt \, .$$

Below we will prove that $E_x|\xi_t| < \infty$, and this along with the boundedness of derivatives of $v$ shows that the process $v_{(\xi_t)}(x_t(x))$ is a martingale on $[0, s]$, and specifically,

$$v_{(\xi)}(x) = E_x v_{(\xi_s)}(x_s) \, .$$

Recall that $s = \tau(x)$ and take $r = \xi^d/2s$. Then $\xi_s^d = 0$ and from the above formula we conclude

$$|v_{(\xi)}(x)| = |E_x g_{\xi_s}(x_s)| \leq \sup_{x^d=0} |g_x| E_x |\xi_s| \, .$$

It remains to estimate the last expectation. To do this we apply Itô's formula to $|\xi_t|^2 e^{-Nt}$ where $N$ is a constant. Then from (4.1.6) we get

$$d|\xi_t|^2 e^{-Nt} = e^{-Nt}\{2(\xi_t, b_{(\xi_t)}) + \|\sigma_{(\xi_t)} + r\sigma\|^2 - N|\xi_t|^2\} \, dt + dm_t$$

with a local martingale $m_t$. If the constant $N$ is large enough (but depends on our initial data as it is stated in the theorem), then it follows that

$$d|\xi_t|^2 e^{-Nt} \leq Nr^2 \, dt + dm_t, \qquad E|\xi_s|^2 e^{-Ns} \leq |\xi|^2 + Nr^2 s \, ,$$

where we have used Lemma 2.1.6. Finally, since $r^2 s = (\xi^d)^2/(4x^d)$ and $E|\xi_t| \leq (E|\xi_t|^2)^{1/2}$, the theorem is proved. $\qquad\square$

REMARK 4.1.4. The result of Theorem 4.1.3 is rather sharp. To show that consider a degenerate diffusion corresponding to the heat equation. Let $d = 2$ and $d_1 = 1$. Define $x_t(x)$ as a solution of the "equation"

$$x_t^1 = x^1 + w_t, \quad x_t^2 = x^2 - t \, .$$

Next, take $D$ from Theorem 4.1.3 and let $g(x) = g(x^1) = |x^1|/(1 + |x^1|^2)$, $v(x) = Eg(x_{\tau(x)}(x))$. Since the first exit time $\tau(x)$ of $x_t(x)$ from $D$ equals $x^2$ (when $x \in D$), we have

$$v(x) = Eg(x^1 + w_{x^2}) = \frac{1}{\sqrt{2\pi x^2}} \int_{-\infty}^{\infty} g(y) e^{-\frac{1}{2x^2}(y-x^1)^2} \, dy \, .$$

The function $v$ satisfies

$$\frac{1}{2} v_{x^1 x^1} - v_{x^2} = 0 \text{ in } D, \quad v = g \text{ on } \partial D \, .$$

Hence its derivative with respect to $x^2$ is one half of its second-order derivative with respect to $x^1$. The latter is a derivative of convolution and we know that we can differentiate either function in convolution. Also notice that the generalized second-order derivative of $g$ equals $g''(x)I_{x\neq 0} + 2\delta_0$, where $\delta_0$ is the $\delta$-function at zero. Therefore,

$$v_{x^2}(0, x^2) = \frac{1}{2\sqrt{2\pi x^2}} \int_{-\infty}^{\infty} g''(y)e^{-\frac{1}{2x^2}y^2} dy + \frac{1}{\sqrt{2\pi x^2}} \sim \frac{1}{\sqrt{2\pi x^2}}$$

as $x^2 \to 0$ and we get exactly the same effect of blow up of the normal derivative and with exactly the same rate: $1/\sqrt{x^d}$ as in Theorem 4.1.3. We can also give now a probabilistic explanation of this blow up as in Remark 4.1.2.

## 4.2. Using time-change related quasiderivatives II. The case of processes non-degenerate along the normal to the boundary

Above we were able to treat Examples 4.1.1 and Theorem 4.1.3 by using time-change related quasiderivatives with *nonrandom* rates. The reason for this was that either we knew in advance where the phase process hits the boundary or what time is needed to reach the boundary. Here we will see that sometimes even if both the hitting point and the exit time could be highly uncertain it is still possible, by using *random* rates, to turn quasiderivatives so that they become tangent to the boundary when the phase process hits it.

EXAMPLE 4.2.1. Let $d = 2$ and let $D = \{(x, y) : x, y \in \mathbb{R}, y > 0\}$. Consider Laplace's equation

(4.2.1) $$\Delta u = 0$$

in $D$ with the boundary data

$$u(x, 0) = g(x) = |x|/(1 + |x|^2).$$

Here, by the Poisson formula

$$u(x, y) = \frac{1}{\pi} \int_{-\infty}^{\infty} g(x + z)\frac{y}{y^2 + z^2} dz$$

$$= -\frac{1}{\pi} \int_{-\infty}^{\infty} g'(x + z) \arctan \frac{z}{y} dz.$$

Hence, as $y \downarrow 0$

$$u_y(0, y) = \frac{1}{\pi} \int_{-\infty}^{\infty} \frac{zg'(z)}{y^2 + z^2} dz = \frac{1}{\pi} \int_{-\infty}^{\infty} \frac{|z|}{(y^2 + z^2)(1 + z^2)} dz$$

$$- \frac{1}{\pi} \int_{-\infty}^{\infty} \frac{2|z|^3}{(y^2 + z^2)(1 + z^2)^2} dz \sim \frac{1}{\pi} \int_{-\infty}^{\infty} \frac{|z|}{(y^2 + z^2)(1 + z^2)} dz$$

$$= \frac{1}{\pi} \int_{0}^{\infty} \frac{1}{(y^2 + z)(1 + z)} dz = \frac{1}{\pi(1 - y^2)} \log \frac{y^2 + z}{1 + z}\Big|_{z=0}^{\infty} = \frac{-2\log y}{\pi(1 - y^2)}.$$

In this section we first show how one can get the result of Example 4.2.1 probabilistically and then extend it to general processes in domains which are nondegenerate along the normal to the boundary.

### 4.2.1. Working out Example 4.2.1 probabilistically

To make notations less ambiguous we consider the multidimensional case and we take $v$ as a bounded solution of equation (4.2.1) in

$$D = \{x \in \mathbb{R}^d : x^d > 0\}$$

with boundary condition

$$v(x) = g(x) \quad \text{for} \quad x^d = 0,$$

where, as everywhere in the chapter, $g$ is a smooth and bounded function on $\mathbb{R}^{d-1}$. From Itô's formula we know that for $x \in D$

$$v(x) = Eg(x + w_{\tau(x)}),$$

where $\tau(x)$ is the first exit time of $x + w_t$ from $D$. We will show how to prove probabilistically that

$$|v_{x^d}(x)| \le N(\sup |g_x| + \sup |g|)|\log x^d| \quad \text{if} \quad x^d > 0.$$

Actually, in this section we will prove only that

(4.2.2) $$|v_{x^d}(x)| \le N(\sup |g_x| + \sup_{x^d=1} |v_x|)|\log x^d| \quad \text{if} \quad x^d \in (0, 1).$$

It is enough since in Section 4.4, we will present a probabilistic proof that derivatives of harmonic functions at a point can be estimated in terms of their maximum values on the boundary of the unit ball with center at this point.

Fix an $x \in D_1 = \{x : x^d \in (0, 1)\}$, $\xi \in \mathbb{R}^d$ and set

$$\tau_1 = \inf\{t \ge 0 : x^d + w_t^d \notin (0, 1)\}.$$

We know that the process

(4.2.3) $$v_{(\xi_t)}(x + w_t)$$

is a local martingale on $[0, \tau_1)$ if

$$\xi_t = \xi + \int_0^t r_s \, dw_s,$$

where $r_s$ is a real-valued process which is adapted and locally square integrable with respect to $t$ with probability one.

We will take the process $r_t$ in such a way that

(4.2.4)
$$E \left( \int_0^{\tau_1} r_s^2 \, ds \right)^{1/2} < \infty .$$

In this case by Davis's inequality and by the fact that the derivatives of $v$ are bounded the process (4.2.3) is uniformly integrable on $[0, \tau_1]$, and therefore

(4.2.5)
$$v_{(\xi)}(x) = E v_{(\xi_{\tau_1})}(x + w_{\tau_1}) .$$

From (4.2.5) and from Davis's inequality it follows that

$$|v_{(\xi)}(x)| \le \sup_{x^d = 1} |v_x| E |\xi_{\tau_1}| + E |v_{(\xi_{\tau_1})}(x + w_{\tau_1})| I_{\tau_1 = \tau(x)}$$

(4.2.6)
$$\le \sup_{x^d = 1} |v_x| \left( |\xi| + 3E \left( \int_0^{\tau_1} r_s^2 \, ds \right)^{1/2} \right)$$
$$+ E |v_{(\xi_{\tau_1})}(x + w_{\tau_1})| I_{\tau_1 = \tau(x)} .$$

We do not want to allow the derivatives of $v$ on $\{x^d = 0\}$ to enter (4.2.6) and since $v = g$ there, we can achieve this if

$$\xi_{\tau_1}^d = 0 \quad \text{(a.s.) on the set} \quad \{\omega : \tau_1 = \tau(x)\} .$$

Therefore, we need to find a process $r_t$ such that inequality (4.2.4) holds along with the equality

(4.2.7)
$$\int_0^{\tau_1} r_t \, dw_t^d = -\xi^d \quad \text{(a.s.) on} \quad \{\omega : \tau_1 = \tau(x)\} .$$

To proceed further we need the following.

LEMMA 4.2.2. *For any adapted, jointly measurable, and locally square integrable in t process $f_t$ and for any stopping time $\gamma$ we have*

(4.2.8)
$$I := E \left( \int_0^{\gamma} f_t^2 \exp \left( 2 \int_0^t f_s \, dw_s^d - \int_0^t f_s^2 \, ds \right) dt \right)^{1/2}$$
$$\le 1 + 4E \int_0^{\gamma} f_t^2 \exp \left( \int_0^t f_s \, dw_s^d - \frac{1}{2} \int_0^t f_s^2 \, ds \right) dt =: 1 + 4J^2 .$$

PROOF. Introduce

$$m_t = \exp \left( \int_0^t f_s \, dw_s^d - \frac{1}{2} \int_0^t f_s^2 \, ds \right) ,$$

$$I_1 = \left( \int_0^{\gamma} f_t^2 \exp \left( \int_0^t f_s \, dw_s^d - \frac{1}{2} \int_0^t f_s^2 \, ds \right) dt \right)^{1/2} .$$

Then the expression under the expectation sign on the left in (4.2.8) is less than

$$I_1 \sup_{t \le \gamma} m_t^{1/2}$$

and by Hölder's inequality

$$I \le E I_1 \sup_{t \le \gamma} m_t^{1/2} \le J \left( E \sup_{t \le \gamma} m_t \right)^{1/2}.$$

On the other hand

$$m_t = 1 + \int_0^t f_s m_s \, dw_s^d,$$

so that by Davis's inequality

$$E \sup_{t \le \gamma} m_t \le 1 + 3E \left( \int_0^\gamma f_t^2 m_t^2 \, dt \right)^{1/2} = 1 + 3I.$$

Hence,

(4.2.9) $$I \le J(1 + 3I)^{1/2}.$$

Furthermore, if $I \le 1$, then (4.2.8) is obvious and if $1 \le I$, then (4.2.9) implies that

$$I \le 2JI^{1/2}, \quad I^{1/2} \le 2J, \quad I \le 4J^2$$

and (4.2.8) holds again. Of course, the last argument requires $I < \infty$, but this point can be easily fixed by appropriate approximating $\gamma$. The lemma is proved. $\square$

Now take a function $\alpha(y)$ on $(0, 1)$ and on $[0, \tau_1)$ solve the equation

(4.2.10) $$\xi_t^d = \xi^d + \int_0^t \alpha(x^d + w_s^d) \xi_s^d \, dw_s^d.$$

Once this equation is solved, we define

$$f_t = \alpha(x^d + w_t^d), \quad r_t = f_t \xi_t^d$$

for $t < \tau_1$, and $f_t = r_t = 0$ otherwise. Observe then that the solution of (4.2.10) is known:

$$\xi_t^d = \xi^d \exp \left( \int_0^t f_s \, dw_s^d - \frac{1}{2} \int_0^t f_s^2 \, ds \right),$$

and that (4.2.7) means that

(4.2.11) $$\exp \left\{ \int_0^{\tau_1} f_s \, dw_s^d - \frac{1}{2} \int_0^{\tau_1} f_s^2 \, ds \right\} = 0 \quad \text{(a.s.)} \quad \text{on} \quad \{\omega : \tau_1 = \tau(x)\}.$$

Consequently, if we denote

(4.2.12) $$\beta(x^d) = E \int_0^{\tau_1} f_t^2 \exp\left(\int_0^t f_s \, dw_s^d - \frac{1}{2}\int_0^t f_s^2 \, ds\right) dt,$$

then by Lemma 4.2.2 and by (4.2.6), under the assumption that condition (4.2.11) is satisfied, the following inequalities hold

$$|v_{(\xi)}(x)| \leq \sup_{x^d=1} |v_x| \left(|\xi| + 3E\left(\int_0^{\tau_1} r_s^2 \, ds\right)^{1/2}\right) + E|g_{(\xi_{\tau_1})}(x + w_{\tau_1})|$$

(4.2.13)
$$\leq (\sup_{x^d=1} |v_x| + \sup |g_x|)\left(|\xi| + 3E\left(\int_0^{\tau_1} r_s^2 \, ds\right)^{1/2}\right)$$

$$\leq N(\sup_{x^d=1} |v_x| + \sup |g_x|)(|\xi| + |\xi^d|\beta(x^d)).$$

Thus, now we need to construct $\alpha$ such that $\beta$ is finite, and at the same time such that condition (4.2.11) is satisfied. Note that the last condition implies in particular that $\alpha(y)$ is singular near zero.

One of possible ideas to find $\alpha$ comes from the fact that $\beta(x^d)$, as a function of $x^d$, should satisfy the corresponding Kolmogorov equation, which is fairly easy to find if one "hides" the exponential martingale in (4.2.12) with the help of the Girsanov theorem. We mean the following equation:

(4.2.14) $$\frac{1}{2}\beta'' - \alpha\beta' + \alpha^2 = 0 \quad \text{on } (0, 1).$$

Actually, if we consider this equation as an equation about $\alpha$ given $\beta$, and we recall that from Example 4.2.1 we know that $\beta(x^d)$ should go to infinity when $x^d \downarrow 0$ at least as fast as $|\log x^d|$, then it is not hard to make a good choice of $\beta$. Namely, we take

$$\beta_0(y) := -2\log\frac{y}{2},$$

and we observe that this function is strictly positive and satisfies equation (4.2.14) for $y \in (0, 1)$ with

$$\alpha(y) = -\frac{1}{y}.$$

At last, we want to justify our argument about $\beta$ from (4.2.12) and about the Kolmogorov equation. As is easy to show by Itô's formula, the process

$$\gamma(t) := \beta_0(x^d + w_{t\wedge\tau(x)}^d) \exp\left(\int_0^{t\wedge\tau(x)} f_s \, dw_s^d - \frac{1}{2}\int_0^{t\wedge\tau(x)} f_s^2 \, ds\right)$$

$$+ \int_0^{t\wedge\tau(x)} f_s^2 \exp\left(\int_0^s f_p \, dw_p^d - \frac{1}{2}\int_0^s f_p^2 \, dp\right) ds$$

is a local martingale. Since it is nonnegative on $[0, \tau_1)$, the process $\gamma(t \wedge \tau_1)$ is a supermartingale and its trajectories are bounded with probability one. But on the set $\{\tau_1 = \tau(x)\}$ we obviously have

$$\lim_{t \uparrow \tau_1} \beta_0(x^d + w^d_{t \wedge \tau(x)}) = \infty,$$

and therefore, condition (4.2.11) is fulfilled. Moreover,

$$\beta_0(x^d) \geq E\gamma(\tau_0) \geq E \int_0^{\tau_0} f_s^2 \exp\left(\int_0^s f_p \, dw_p^d - \frac{1}{2} \int_0^s f_p^2 \, dp\right) ds = \beta(x^d),$$

which means that in (4.2.13) we can replace $\beta(x^d)$ by $\beta_0(x^d)$, and this leads immediately to inequality (4.2.2).

REMARK 4.2.3. Equation (4.2.14), as an equation about $\alpha$ given $\beta$, has a solution only if

$$2\beta'' \leq (\beta')^2.$$

It turns out that there is no function $\beta$ satisfying the last inequality on $[0, \infty)$ and having an infinite limit at zero. This is the reason why we considered (4.2.14) on $(0, 1)$ only, and why we confined ourselves to estimate (4.2.2).

REMARK 4.2.4. Example 4.2.1 shows that estimate (4.2.2) is sharp in what concerns the rate of blow up of the normal derivative of $v$. We can also now give a probabilistic explanation of this blow up as in Remark 4.1.2.

### 4.2.2. General case. Adding parameters $P$ into the picture

In this subsection we consider general equation (3.0.1) and along with time-change related quasiderivatives also use the quasiderivatives associated with *the parameters $P$* arising while changing the driving Wiener process.
    Set

$$a = (1/2)\sigma\sigma^*.$$

Assume that $D$ is bounded

(4.2.15) $$D \subset \mathbb{R}^d, \quad D \in C^3,$$

and that

$$P_x(\tau < \infty) = 1, \quad \forall x \in D.$$

By $n(y)$ we denote the unit inward normal vector to $\partial D$ at each $y \in \partial D$. We also need a function $\psi$ such that

$$\psi \in C^3, \quad \psi > 0 \quad \text{in} \quad D, \quad \psi = 0 \quad \text{and} \quad |\psi_x| \geq 1 \quad \text{on} \quad \partial D.$$

Due to (4.2.15) such functions always exist.

THEOREM 4.2.5. *Recall that the function*

$$v(x) := E_x g(x_\tau),$$

*is assumed to be continuously differentiable in $\bar{D}$. Also assume that*

(4.2.16)                          $(an, n) > 0$   *on*   $\partial D$.

*Then there is an $\varepsilon > 0$ such that for any $\xi \in \mathbb{R}^d$ in $D$*

(4.2.17)                $|v_{(\xi)}| \leq N(|\xi| + |\psi_{(\xi)}| \log \psi)(\sup_{\partial D} |g_x| + \sup_{\psi \geq \varepsilon} |v_x|),$

*where $N$ is a constant independent of $\xi$.*

REMARK 4.2.6. The presence of

$$\sup_{\psi \geq \varepsilon} |v_x|$$

on the right looks unpleasant. In this connection note that it is shown in [9] that (4.2.17) holds in the whole of $D$ with no such term on the right and without any apriori smoothness assumption on $v$. Instead, some almost necessary conditions are imposed on $\sigma$ and $b$ guaranteeing that the moments of quasiderivatives do not grow too fast in $t$.

It is also worth noting that at the moment no analytic proof even of Theorem 4.2.5 is known.

Before proving the theorem we prove two auxiliary results.

LEMMA 4.2.7. *There exist a function $\bar{\psi}$ on $\mathbb{R}^d$ and an $\varepsilon > 0$ such that for*

$$D' := \{x \in D : 0 < \bar{\psi} < \varepsilon\}$$

*it holds that $\partial D \subset \partial D'$,*

(4.2.18)        $\bar{\psi} \in C^3$,   $\bar{\psi} = 0$   *and*   $|\bar{\psi}_x| \geq 1$   *on*   $\partial D$,

(4.2.19)                $L\bar{\psi} \leq -1$   *and*   $\sum_k (\bar{\psi}_{(\sigma^k)})^2 \geq 1$   *in*   $D'$,

*where*

$$Lu := a^{ij} u_{x^i x^j} + b^i u_{x^i}.$$

PROOF. We are going to prove that for an appropriate big constant $c > 0$

$$\bar{\psi} = c\psi(1 - c\psi), \quad \varepsilon = 1/4$$

possess the required properties. Of course (4.2.18) holds for any $c \geq 1$.

Next, notice that

$$D' \subset \{x \in D : \psi < 1/(4c)\},$$

on $\partial D$ it holds that $\psi_x = |\psi_x| n$, and hence

$$\sum_k (\psi_{(\sigma^k)})^2 = 2(a\psi_x, \psi_x) = 2|\psi_x|(an, n) \geq 2\delta \quad \text{on} \quad \partial D,$$

where $\delta > 0$ is a constant. By continuity

$$\sum_k (\psi_{(\sigma^k)})^2 = 2(a\psi_x, \psi_x) \geq \delta \quad \text{in} \quad D'$$

if $c$ is large enough.

It follows that in $D'$ for large $c$

$$L\bar{\psi} = cL\psi - c^2(2\psi L\psi + 2(a\psi_x, \psi_x)) \leq cN - c^2\delta \leq -1,$$
$$\bar{\psi}_{(\sigma^k)} = c(1 - 2c\psi)\psi_{(\sigma^k)}, \quad \sum_k (\bar{\psi}_{(\sigma^k)})^2 = c^2(1 - 2c\psi)^2 2(a\psi_x, \psi_x) \geq 1.$$

The lemma is proved. $\qquad\square$

LEMMA 4.2.8. *Let $\xi_t$ be an $\mathbb{R}^d$-valued process having stochastic differential*

$$d\xi_t = \beta_t^k \, dw_t^k + \alpha_t \, dt.$$

*Assume that $|\beta_t^k| \leq \gamma_t |\xi_t|$, where $\gamma_t^2$ is locally integrable in $t$ (a.s.). Then*

$$d|\xi_t| = |\xi_t|^{-1}(\xi_t, \beta_t^k) \, dw_t^k$$
$$+ \frac{1}{2} \left[ 2|\xi_t|^{-1}(\xi_t, \alpha_t) + |\xi_t|^{-1} \sum_k |\beta_t^k|^2 - |\xi_t|^{-3} \sum_k (\xi_t, \beta_t^k)^2 \right] dt,$$

*where $0^{-1}0 := 0$ if such expressions occur. In particular, if $d = 1$ so that $\xi_t$ is real valued, then*

$$d|\xi_t| = \beta_t^k \operatorname{sign} \xi_t \, dw_t^k + \alpha_t \operatorname{sign} \xi_t \, dt.$$

PROOF. By Itô's formula for any constant $\kappa > 0$

$$d\left[|\xi_t|^2 + \kappa\right]^{1/2} = \left[|\xi_t|^2 + \kappa\right]^{-1/2}(\xi_t, \beta_t^k) \, dw_t^k$$
$$+ \frac{1}{2} \left[ 2\left[|\xi_t|^2 + \kappa\right]^{-1/2}(\xi_t, \alpha_t) + \left[|\xi_t|^2 + \kappa\right]^{-1/2} \sum_k |\beta_t^k|^2 \right.$$
$$\left. - \left[|\xi_t|^2 + \kappa\right]^{-3/2} \sum_k (\xi_t, \beta_t^k)^2 \right] dt.$$

We write this equation in the integral form and pass to the limit as $\kappa \downarrow 0$. In the integral with respect to $t$ we use the dominated convergence theorem and the fact that $\xi_t$ is continuous so that its trajectories are bounded on each finite time interval. This and the inequalities

$$[|\xi_t|^2 + \kappa]^{-1/2} |(\xi_t, \alpha_t)| \leq |\alpha_t|,$$

$$[|\xi_t|^2 + \kappa]^{-1/2} \sum_k |\beta_t^k|^2 + [|\xi_t|^2 + \kappa]^{-3/2} \sum_k (\xi_t, \beta_t^k)^2 \leq N |\xi_t| \gamma_t^2$$

take care of the integral against $dt$.

To pass to the limit in the stochastic integral it suffices to rely on the equivalence of convergences to zero of stochastic integrals and their quadratic variations and the estimate

$$\left| [|\xi_t|^2 + \kappa]^{-1/2} (\xi_t, \beta_t^k) \right|^2 \leq |\xi_t| \gamma_t^2.$$

The lemma is proved.                                                                    □

PROOF OF THEOREM 4.2.5. Lemma 4.2.7 shows that without losing generality we may assume that $\psi$ itself possesses the properties of this lemma for an $\varepsilon > 0$. Then introduce

$$A := \sum_k (\psi_{(\sigma^k)})^2, \quad \rho = -\frac{1}{A} \psi_{(\sigma^k)} (\psi_{(\sigma^k)})_{(\xi)}, \quad r = \rho + \frac{\psi_{(\xi)}}{\psi},$$

$$P^{ik} = \frac{1}{A} [\psi_{(\sigma^k)} (\psi_{(\sigma^i)})_{(\xi)} - \psi_{(\sigma^i)} (\psi_{(\sigma^k)})_{(\xi)}], \quad i, k = 1, \ldots, d_1, \quad P = (P^{ik}).$$

The functions $\rho(x, \xi), r(x, \xi)$, and $P(x, \xi)$ are linear in $\xi$ but $r(x, \xi)$ is not bounded in $D'$ for fixed $\xi$. Therefore, for $\delta \in (0, \varepsilon)$ we set

$$D_\delta' = \{x \in D' : \psi > \delta\}, \quad (\rho_\delta, r_\delta, P_\delta)(x, \xi) = (\rho, r, P)(x, \xi) I_{D_\delta'}(x).$$

Observe that for $x \in D \setminus \bar{D}'$ estimate (4.2.17) is obvious. Therefore, we may concentrate on proving (4.2.17) in $D'$. Fix $x_0 \in D'$, $\xi_0 \in \mathbb{R}^d$, and $\delta \in (0, \varepsilon)$ and define $\xi_t$ as a solution of

$$(4.2.20) \qquad d\xi_t = [\sigma_{(\xi_t)} + r_\delta \sigma + \sigma P_\delta] \, dw_t + [b_{(\xi_t)} + 2 r_\delta b] \, dt, \quad t \geq 0,$$

with initial value $\xi_0$, where in the argument of the coefficients including $r_\delta$ and $P_\delta$ we omit $\xi_t$ and $x_t := x_t(x_0)$ for brevity. We will also write $\tau$ instead of $\tau(x_0)$.

As we know from Theorem 3.2.1 the process $\xi_t$ is a quasiderivative and therefore

$$v_{(\xi_t)}(x_t)$$

is a local martingale on $[0, \tau)$. Also by assumption $v_x$ is bounded in $D'$ and all moments of $\xi_t$ exist since $\xi_t$ is a solution of linear equation with bounded coefficients. Therefore,

$$v_{(\xi_{t \wedge \tau})}(x_{t \wedge \tau})$$

is a martingale and

(4.2.21) $$v_{(\xi_0)}(x_0) = E_{x_0} v_{(\xi_{\tau_\delta})}(x_{\tau_\delta}),$$

where

$$\tau_\delta = \inf\{t \geq 0 : x_t \notin D'_\delta\}.$$

We now want to show that $\xi_{\tau_\delta}$ is almost tangential to $\partial D'_\delta$ if $x_t$ hits its boundary where $\psi = \delta$. To do this use Itô's formula to find that for $t \leq \tau_\delta$

$$d\psi_{(\xi_t)} = [(L\psi)_{(\xi_t)} + 2rL\psi] dt + [(\psi_{(\sigma^i)})_{(\xi_t)} + r\psi_{(\sigma^i)} + \psi_{(\sigma^k)} P^{ki}] dw_t^i,$$

where again we drop the obvious values of the arguments. A remarkable fact about this equation is that owing to our choice of $r$ and $P$

$$(\psi_{(\sigma^i)})_{(\xi_t)} + r\psi_{(\sigma^i)} + \psi_{(\sigma^k)} P^{ki} = \frac{1}{\psi} \psi_{(\sigma^i)} \psi_{(\xi_t)},$$

which is proportional to $\psi_{(\xi_t)}$. By Lemma 4.2.8 for $t \leq \tau_\delta$

$$d|\psi_{(\xi_t)}| = [(L\psi)_{(\xi_t)} + 2rL\psi] \operatorname{sign} \psi_{(\xi_t)} dt + \frac{1}{\psi} \psi_{(\sigma^i)} |\psi_{(\xi_t)}| dw_t^i.$$

We also use Lemma 4.2.8 to find $d|\xi_t|$ and for

$$B(x, \xi) := \sqrt{\varepsilon}(1 + \sqrt{\psi(x)})|\xi| - |\psi_{(\xi)}(x)| \log \psi(x)$$

and $t \leq \tau_\delta$ we obtain

$$dB(x_t, \xi_t) = \Gamma(x_t, \xi_t) dt + dm_t,$$

where $m_t$ is a local martingale on $[0, \tau_\delta)$ and

$$\Gamma(x, \xi) = \frac{1}{2}\sqrt{\varepsilon}(1 + \sqrt{\psi}) \left[ \frac{2}{|\xi|}(\xi, \bar{b}) + \frac{1}{|\xi|} \sum_k |\bar{\sigma}^k|^2 - \frac{1}{|\xi|^3} \sum_k (\xi, \bar{\sigma}^k)^2 \right]$$

$$+ \frac{1}{2} \frac{\sqrt{\varepsilon}}{|\xi|\sqrt{\psi}} (\bar{\sigma}^k, \xi) \psi_{(\sigma^k)} - \frac{1}{\psi^2} |\psi_{(\xi)}| (\psi(1 + 2\log\psi)L\psi + A/2)$$

$$+ \frac{|\xi|}{2\sqrt{\psi}} \sqrt{\varepsilon} L\psi - \frac{|\xi|}{8\psi^{3/2}} \sqrt{\varepsilon} A - \log\psi (2\rho L\psi + (L\psi)_{(\xi)}) \operatorname{sign} \psi_{(\xi)},$$

with

$$\bar{b} := b_{(\xi)} + 2rb, \quad \bar{\sigma} := \sigma_{(\xi)} + r\sigma + \sigma P.$$

It is not hard to see that with constants $N$ independent of $\varepsilon$ we have in $D'$ that

$$|\bar{\sigma}| \le N\left(|\xi| + \frac{|\psi_{(\xi)}|}{\psi}\right), \quad \frac{|\bar{\sigma}|^2}{|\xi|} \le N\left(|\xi| + \frac{|\psi_{(\xi)}|}{\psi^2}\right),$$

$$\frac{2}{|\xi|}|(\xi, \bar{b})| \le N\left(|\xi| + \frac{|\psi_{(\xi)}|}{\psi}\right) \le N\frac{\varepsilon^{3/2}}{\psi^{3/2}}|\xi| + N\frac{|\psi_{(\xi)}|}{\psi^2},$$

$$\frac{1}{2}\frac{\sqrt{\varepsilon}}{|\xi|\sqrt{\psi}}(\bar{\sigma}^k, \xi)\psi_{(\sigma^k)} \le \frac{1}{2}\frac{\sqrt{\varepsilon}}{\sqrt{\psi}}|\bar{\sigma}^k|\,|\psi_{(\sigma^k)}|$$

$$\le N\frac{\sqrt{\varepsilon}}{\sqrt{\psi}}\left(|\xi| + \frac{|\psi_{(\xi)}|}{\psi}\right) \le N\frac{\varepsilon^{3/2}}{\psi^{3/2}}|\xi| + N\varepsilon\frac{|\psi_{(\xi)}|}{\psi^2},$$

$$|\log\psi(2\rho L\psi + (L\psi)_{(\xi)})| \le N|\xi|\,|\log\psi| \le N\frac{\varepsilon}{\psi^{3/2}}|\xi|,$$

$$-\frac{1}{\psi^2}|\psi_{(\xi)}|(\psi(1 + 2\log\psi)L\psi + A/2) \le -\frac{1}{4\psi^2}|\psi_{(\xi)}|,$$

where the last two estimates hold if $\varepsilon$ is small enough due to the fact that $A \ge 1$. Upon collecting our estimates we see that

$$\Gamma \le N\frac{\varepsilon}{\psi^{3/2}}|\xi| + N\frac{\sqrt{\varepsilon}}{\psi^2}|\psi_{(\xi)}| - \frac{\sqrt{\varepsilon}}{8\psi^{3/2}}|\xi| - \frac{1}{4\psi^2}|\psi_{(\xi)}|.$$

Hence $\Gamma(x, \xi) \le 0$ for $x \in D'$ if $\varepsilon$ is small enough. We choose and fix an appropriate $\varepsilon$. Then $B(x_t, \xi_t)$ is a local supermartingale on $[0, \tau_\delta]$ and since it is nonnegative, it is just a supermartingale and, in particular, its trajectories are bounded on $[0, \tau_\delta)$. The latter, of course, implies that $|\psi_{\xi_t}(x_t)\log\psi(x_t)|$ is bounded and consequently $\psi_{\xi_t}(x_t)$ becomes small and $\xi_t$ becomes close to a tangent plane to $\{\psi = \delta\}$ if $x_t$ approaches $\{\psi = \delta\}$. We see that we are able to turn the quasiderivative the way we wanted.

To be more precise observe that (4.2.21) yields

$$v_{(\xi_0)}(x_0) \le \sup_{x \in \partial D'_\delta} \sup_{\xi \ne 0} \frac{|v_{(\xi)}(x)|}{B(x, \xi)} E_{x_0} B(x_{\tau_\delta}, \xi_{\tau_\delta})$$

(4.2.22)

$$\le B(x_0, \xi_0) \sup_{x \in \partial D'_\delta} \sup_{|\xi| = 1} \frac{|v_{(\xi)}(x)|}{B(x, \xi)} =: B(x_0, \xi_0)I_\delta.$$

Here

$$I_\delta \le \sup_{\psi = \varepsilon} \sup_{|\xi| = 1} \frac{|v_{(\xi)}(x)|}{B(x, \xi)} + \frac{|v_{(\xi(\delta))}(x(\delta))|}{B(x(\delta), \xi(\delta))} \le N\sup_{\psi = \varepsilon}|v_x| + \frac{|v_{(\xi(\delta))}(x(\delta))|}{B(x(\delta), \xi(\delta))},$$

where $x(\delta) \in D'$ and $\xi(\delta)$ are some points such that $\psi(x(\delta)) = \delta$ and $|\xi(\delta)| = 1$.

A subsequence of $x(\delta), \xi(\delta)$ converges to some $y, \eta$, such that $y \in \partial D$ and $|\eta| = 1$. If $\psi_{(\eta)}(y) \neq 0$, then $B(x(\delta), \xi(\delta)) \to \infty$ as $\delta \downarrow 0$ and passing to the limit in (4.2.22) proves that

$$v_{(\xi_0)}(x_0) \leq NB(x_0, \xi_0) \sup_{\psi = \varepsilon} |v_x|$$

(4.2.23)

$$\leq N(|\xi_0| + |\psi_{(\xi_0)} \log \psi(x_0)|)(\sup_{\partial D} |g_x| + \sup_{\psi \geq \varepsilon} |v_x|).$$

The same is true if $\psi_{(\eta)}(y) = 0$ since then

$$\overline{\lim_{\delta \downarrow 0}} \frac{|v_{(\xi(\delta))}(x(\delta))|}{B(x(\delta), \xi(\delta))} \leq \overline{\lim_{\delta \downarrow 0}} \frac{|v_{(\xi(\delta))}(x(\delta))|}{\sqrt{\varepsilon}(1 + \sqrt{\psi(x(\delta))})} = \frac{|g_{(\eta)}(y)|}{\sqrt{\varepsilon}} \leq \varepsilon^{-1/2} |g_x(y)|.$$

We have proved (4.2.23) which is a one sided version of (4.2.17). Now it only remains to take $-\xi_0$ in place of $\xi_0$. The theorem is proved.

## 4.3. Example 4.1.1 revisited using general quasiderivatives

The methods in Example 4.1.1 leading to an estimate of $v_{x^2}(x)$ work only if $x$ is on the $x^2$-axis. In particular, the presented approach based on quasiderivatives does not work since there is no constant $r$ such that

$$rw_{\tau(x)}(x^1 + w_{\tau(x)}) + x^2 = 0.$$

Let us try to deal with arbitrary $x \in D$ by using more general quasiderivatives this time for any $\xi \in \mathbb{R}^2$. Generally, we have the following quasiderivatives

$$d\xi_t^1 = r_t \, dw_t - \pi_t \, dt, \quad d\xi_t^2 = 0.$$

The condition $(\xi_{\tau(x)}, x_{\tau(x)}(x)) = 0$ means that

$$\xi_{\tau(x)}^1 x_{\tau(x)}^1(x) + \xi^2 x^2 = 0, \quad \xi_{\tau(x)}^1 = -\frac{\xi^2 x^2}{x_{\tau(x)}^1(x)},$$

$$\xi^1 + \int_0^{\tau(x)} r_t \, dw_t + \int_0^{\tau(x)} \pi_t \, dt = \frac{\xi^2 x^2}{x_{\tau(x)}^1(x)}.$$

The left-hand side looks like the result of application of Itô's formula to a function $G(x_t(x))$ such that

(4.3.1) $$G(x) = \xi^1, \quad G(y) = \frac{\xi^2 x^2}{y^1}, \quad y \in \partial D.$$

Obviously there are plenty of such functions.

If we have one of them then we take

$$r_t = G_{x^1}(x_t(x)), \quad \pi_t = -(1/2)G_{x^1x^1}(x_t(x))$$

and we obtain

$$v_{(\xi)}(x) = E_x g_{(\xi_\tau)}(x_\tau) + E_x g(x_\tau) \int_0^\tau \pi_t \, dw_t =: I + J,$$

where

$$I = E_x g_{x^1}(x_\tau)\xi_\tau^1 = \xi^2 x^2 E_x g_{x^1}(x_\tau)(x_\tau^1)^{-1},$$
$$|I| \le |\xi^2 x^2| \, |g|_1 E_x |x_\tau^1|^{-1} = |\xi^2 x^2| \, [g]_1 (1 - |x^2|^2)^{-1/2}.$$

Observe that expression $I$ is totally independent of the choice of $G$. To estimate $J$ we choose $G$ so that it is quadratic on each horizontal line, that is

$$G(y) = \frac{1 - |x^2|^2 - (y^1)^2}{1 - |x|^2}\left(\xi^1 - \frac{\xi^2 x^2 x^1}{1 - |x^2|^2}\right) + \frac{\xi^2 x^2 y^1}{1 - |x^2|^2}.$$

Then $G_{x^1x^1}$ is a constant,

$$J = E_x[g(x_\tau) - g(x)] \int_0^\tau \pi_t \, dw_t$$
$$= -(1/2)G_{x^1x^1} E_x[g(x_\tau) - g(x)] \int_0^\tau dw_t$$
$$= -(1/2)G_{x^1x^1} E_x[g(x_\tau) - g(x)][x_\tau - x],$$

so that

$$|J| \le \frac{1}{1 - |x|^2}\left|\xi^1 - \frac{\xi^2 x^2 x^1}{1 - |x^2|^2}\right| [g]_1 E_x |x_\tau^1 - x^1|^2,$$

where

$$E_x|x_\tau^1 - x^1|^2 = E_x|x_\tau^1|^2 - |x^1|^2 = 1 - |x|^2.$$

Finally after noticing that $|x^1| \le \sqrt{1 - |x^2|^2}$ we infer that

$$|v_{(\xi)}(x)| \le [g]_1\left(|\xi^1| + \frac{2|\xi^2 x^2|}{\sqrt{1 - |x^2|^2}}\right),$$

$$|v_{x^1}(x)| \le [g]_1, \quad |v_{x^2}(x)| \le 2[g]_1 \frac{|x^2|}{\sqrt{1 - |x^2|^2}},$$

and these estimates are in an extremely good agreement with (4.1.1).

## 4.4. Using measure-change related quasiderivatives in the case of uniformly nondegenerate processes

### 4.4.1. Laplace's equations

First we consider a classical situation.

EXAMPLE 4.4.1. Let $D = \{x \in \mathbb{R}^d : |x| < 1\}$, $g$ a Borel bounded function on $\partial D$, and for $x \in D$ let $\tau = \tau(x)$ be the first exit time of

$$x_t := x + w_t$$

from $D$, where $w_t$ is a $d$-dimensional Wiener process. Recall that

$$v(x) = E_x g(x_\tau) .$$

As is well known $v$ is infinitely differentiable in $D$ and satisfies

$$(4.4.1) \qquad \Delta v = 0 \quad \text{in} \quad D .$$

From the explicit representation of $v$ by means of the Poisson formula it follows that there exists a constant $N$, depending only on $d$, such that

$$(4.4.2) \qquad |v_x(0)| \leq N \sup_{\partial D} |g| .$$

We want to show that this result can be easily obtained probabilistically by using quasiderivatives without referring to equation (4.4.1). However, as we agreed in the introduction to the chapter we only do this for smooth $g$ and $v$.

We fix $\xi \in \mathbb{R}^d$ and take an adapted $\mathbb{R}^d$-measurable process $\pi_t$ such that

$$(4.4.3) \qquad I := E \int_0^{\tau_0} |\pi_t|^2 \, dt < \infty ,$$

where $\tau_0 = \tau(0)$. Introduce

$$(4.4.4) \qquad \xi_t = \xi - \int_0^t \pi_s \, ds .$$

Then by Theorem 3.2.1 the process

$$\eta_t := v_{(\xi_t)}(w_t) + v(w_t) \int_0^t \pi_s \, dw_s$$

is a local martingale on $[0, \tau_0)$. Condition (4.4.3) implies that

$$E \sup_{t \leq \tau_0} |\xi_t| \leq |\xi| + E \int_0^{\tau_0} |\pi_t| \, dt \leq |\xi| + I^{1/2} (E\tau_0)^{1/2} < \infty .$$

We see that $\eta_{t \wedge \tau_0}$ is uniformly integrable, so that by Lemma 2.4.8

$$v_{(\xi)} = E\eta_0 = E\left[v_{(\xi_{\tau_0})}(w_{\tau_0}) + v(w_{\tau_0}) \int_0^{\tau_0} \pi_t \, dw_t\right]$$

$$= E\left[v_{(\xi_{\tau_0})}(w_{\tau_0}) + g(w_{\tau_0}) \int_0^{\tau_0} \pi_t \, dw_t\right]$$

$$\leq |g|_0 \left(E \int_0^{\tau_0} |\pi_t|^2 \, dt\right)^{1/2} + Ev_{(\xi_{\tau_0})}(w_{\tau_0}).$$

We see that to get (4.4.2) it only remains to choose $\pi_t$ satisfying (4.4.3) so that

(4.4.5)             $$\xi_{\tau_0} = \xi - \int_0^{\tau_0} \pi_t \, dt = 0.$$

Let us take a positive function $h$ on $B$ and look for $\pi_t$ in the form $h(w_t)\xi_t$ for $t \leq \tau_0$ and set $\pi_t = 0$ for $t > \tau_0$. In that case $\xi_t$ satisfies

(4.4.6)             $$\xi_t = \xi - \int_0^t h(w_s)\xi_s \, ds \quad t \leq \tau_0.$$

The solution of (4.4.6) is obvious:

$$\xi_t = \xi \exp\left(-\int_0^t h(w_s) \, ds\right),$$

and to satisfy (4.4.3) and (4.4.5) we need only find a function $h$ such that for $x = 0$ it holds that

(4.4.7)
$$\int_0^{\tau(x)} h(x + w_t) \, dt = \infty \quad (\text{a.s.}),$$

$$u(x) := E \int_0^{\tau(x)} h^2(x + w_t) \exp\left(-2 \int_0^t h(x + w_s) \, ds\right) dt < \infty.$$

Observe that the function $u$ should satisfy the corresponding Kolmogorov equation

$$\frac{1}{2}\Delta u - 2hu + h^2 = 0 \quad \text{in} \quad D.$$

Since we only need an estimate of $u$, inspired by the maximum principle, we will try to find a nonnegative function $u_0$ satisfying the inequality

$$\frac{1}{2}\Delta u_0 - 2hu_0 + h^2 \leq 0 \quad \text{in} \quad D.$$

Of course, to satisfy the first equation in (4.4.7) we need $h$ to blow up near $\partial D$. Thus, we take a constant $C > 0$ and set

(4.4.8)             $$h(x) = \frac{C}{(1 - |x|^2)^2}, \quad u_0(x) = \frac{1}{(1 - |x|^2)^2}.$$

It is easy to check that if the constant $C$ is large enough, we have in $D$ that

$$\frac{1}{2}\Delta u_0 - 2hu_0 \leq -\frac{1}{C^2}h^2 .$$

Finally, we check that this choice of $h$ is right. By Itô's formula applied to

$$u_0(w_t)\exp\left(-2\int_0^t h(w_s)\,ds\right) ,$$

for any $\rho \in (0, 1)$ and $\gamma_\rho = \inf\{t \geq 0 : |w_t| \geq \rho\}$ we get

$$
\begin{aligned}
u_0(0) = {}& E\left[u(w_{\gamma_\rho})\exp\left(-2\int_0^{\gamma_\rho} h(w_s)\,ds\right)\right. \\
& \left. + \int_0^{\gamma_\rho}\left(2hu - \frac{1}{2}\Delta u\right)(w_t)\exp\left(-2\int_0^t h(w_s)\,ds\right)dt\right] \\
\geq {}& \frac{1}{(1-\rho^2)^2}E\exp\left(-2\int_0^{\gamma_\rho} h(w_s)\,ds\right) \\
& + \frac{1}{C^2}E\int_0^{\gamma_\rho} h^2(w_t)\exp\left(-2\int_0^t h(w_s)\,ds\right)dt .
\end{aligned}
$$

By letting here $\rho \uparrow 1$ we immediately obtain both relations in (4.4.7) along with the inequality $u_0(0) \geq u(0)$.

REMARK 4.4.2. Example 4.4.1 shows that if in Example 4.1.1 we add a diffusion with constant $\delta > 0$ in the vertical direction then we have a uniformly nondegenerate process and quasiderivatives can be successfully used to get estimates of derivatives of $v$. One can also apply the result of Subsection 4.2.2. Yet then not only do we use quasiderivatives differently but also these estimates are *not* uniform with respect to $\delta \in (0, 1]$ and deteriorate as $\delta \downarrow 0$.

Generally, even in the case of diffusion with constant coefficients at this moment in the lectures we still do not have *universal* methods allowing us to *steer* $\xi_t$ into the tangent plane. In the last part of the lectures a different idea is exploited. Observe that the linear combination of quasiderivatives with nonrandom coefficients is also a quasiderivative. Our idea in Chapter 5 is based on the fact that sometimes one can allow the coefficients of the linear combination to be random and depend on the future.

REMARK 4.4.3. Example 4.4.1 taken from [9] was further generalized by Thalmaier [18] for Riemannian manifolds and the Laplace-Beltrami operators. The generalization in Subsection 4.4.2 is not covered by [18].

### 4.4.2. General elliptic equations

Now we generalize Example 4.4.1 to obtain the following classical result from the theory of PDE which holds under only continuity condition on the coefficients.

THEOREM 4.4.4. *Let* $D = \{x \in \mathbb{R}^d : |x| < 1\}$ *and assume that a is uniformly nondegenerate. Then*

(4.4.9)
$$|v_x(0)| \le N|g|_{0,\partial D},$$

*where N depends only on bounds of a, b, and their first derivatives and the constant bounding away from zero the eigenvalues of a in D.*

PROOF. As in Example 4.4.1 we fix $\xi \in \mathbb{R}^d$ and take measure-change related quasiderivatives defined by

(4.4.10)
$$d\xi_t = \sigma^k_{(\xi_t)}(x_t)\,dw^k_t + [b_{(\xi_t)}(x_t) - \sigma^k(x_t)\pi^k_t]\,dt, \quad \xi_0 = \xi,$$

where $x_t = x_t(0)$, and try to find the processes $\pi^k_t$ so that

$$\xi_\tau = 0 \quad \text{(a.s.) with} \quad \tau = \tau(0).$$

For simplicity of notation we drop the argument $x_t$ on few occasions below. Since

$$d|\xi_t|^2 = [2(\xi_t, b_{(\xi_t)}) + \|\sigma_{(\xi_t)}\|^2 - 2(\xi_t, \sigma^k)\pi^k_t]\,dt + 2(\xi_t, \sigma^k_{(\xi_t)})\,dw^k_t$$

and $a = (1/2)\sigma\sigma^*$ is uniformly nondegenerate, it is natural to take

(4.4.11)
$$\pi^k_t = h(x_t)(\xi_t, \sigma^k),$$

for an appropriate function $h \ge 0$, which is locally bounded in $D$. Indeed, in that case $\xi_t$ is a quasiderivative on $[0, \tau)$ and

$$2(\xi_t, \sigma^k)\pi^k_t = h(a\xi_t, \xi_t) \ge \varepsilon h|\xi_t|^2,$$

where $\varepsilon > 0$ is a constant under control. Furthermore, on $[0, \tau)$

$$2(\xi_t, b_{(\xi_t)}) + \|\sigma_{(\xi_t)}\|^2 - 2(\xi_t, \sigma^k)\pi^k_t \le -(\varepsilon h - N_1)|\xi_t|^2.$$

For

$$4h_1 := \varepsilon h - N_1, \quad \psi_t := \exp\left(2\int_0^t h_1(x_s)\,ds\right)$$

it follows that on $[0, \tau)$

$$d\left[|\xi_t|^2\psi_t + 2\int_0^t |\xi_s|^2 h_1(x_s)\psi_s\,ds\right]$$
$$= [2(\xi_t, b_{(\xi_t)}) + \|\sigma_{(\xi_t)}\|^2 - 2(\xi_t, \sigma^k)\pi^k_t$$
$$+ 4h_1|\xi_t|^2]\psi_t\,dt + 2(\xi_t, \sigma^k_{(\xi_t)})\psi_t\,dw^k_t \le 2(\xi_t, \sigma^k_{(\xi_t)})\psi_t\,dw^k_t.$$

Thus on $[0, \tau)$

(4.4.12) $\qquad |\xi_t|^2 \psi_t + 2 \int_0^t |\xi_s|^2 h_1(x_s) \psi_s \, ds \leq |\xi|^2 + 2 \int_0^t (\xi_s, \sigma^k_{(\xi_s)}) \psi_s \, dw^k_s .$

Since the left-hand side is nonnegative by Lemma 2.1.6, for any stopping time $\gamma < \tau$ we have

(4.4.13) $\qquad E_0 |\xi_\gamma|^2 \psi_\gamma \leq |\xi|^2, \quad E_0 \int_0^\tau |\xi_s|^2 h_1(x_s) \psi_s \, ds \leq |\xi|^2 .$

Furthermore, for $\gamma$ such that $E_0 \sup_{t \leq \gamma} |\xi_t|^2 \psi_t < \infty$ we find from (4.4.12) and Davis's inequality that

$$E_0 \sup_{t \leq \gamma} |\xi_t|^2 \psi_t \leq |\xi|^2 + N_2 E_0 \left[ \int_0^\gamma |\xi_s|^4 \psi_s^2 \, ds \right]^{1/2} ,$$

where

$$E_0 \left[ \int_0^\gamma |\xi_s|^4 \psi_s^2 \, ds \right]^{1/2} \leq E_0 \left[ \int_0^\gamma |\xi_s|^2 \psi_s \, ds \right]^{1/2} \sup_{t \leq \gamma} |\xi_t| \psi_t^{1/2}$$

$$\leq \frac{1}{2N_2} E_0 \sup_{t \leq \gamma} |\xi_t|^2 \psi_t + 2N_2 E_0 \int_0^\tau |\xi_s|^2 \psi_s \, ds .$$

This and (4.4.13) yield

(4.4.14) $\qquad E_0 \sup_{t \leq \gamma} |\xi_t|^2 \psi_t \leq (4N_2^2 + 1)|\xi|^2$

for $\gamma$ described above provided that $h_1 \geq 1$. By Fatou's lemma (4.4.14) holds for any stopping time $\gamma < \tau$.

Now we choose $h_1$ as in (4.4.8), that is we set

$$h_1(x) = \frac{C}{(1 - |x|^2)^2}, \quad u(x) = \frac{1}{(1 - |x|^2)^2}$$

and observe that as before, if $C$ is large enough, then

$$a^{ij} u_{x^i x^j} + b^i u_{x^i} - 2h_1 u \leq -\frac{1}{C^2} h_1^2 \quad \text{in} \quad D .$$

As before, we denote by $\gamma_\rho$ the first exit time of $x_t$ from $\{x : |x| < \rho\}$ and for $\rho \in (0, 1)$ obtain by Itô's formula that

(4.4.15) $\qquad u(0) \geq \frac{1}{(1 - \rho^2)^2} E_0 \psi_{\gamma_\rho}^{-1} + \frac{1}{C^2} E_0 \int_0^{\gamma_\rho} h_1^2(x_t) \psi_t^{-1} \, dt .$

It follows from here and (4.4.13) that

$$(E_0|\xi_{\gamma_\rho}|)^2 \le E_0|\xi_{\gamma_\rho}|^2\psi_{\gamma_\rho} E_0\psi_{\gamma_\rho}^{-1} \le (1-\rho^2)^2|\xi|^2.$$

Furthermore, for the adjoint process

$$\xi_t^0 = \int_0^t \pi_s \, dw_s$$

owing to Davis's inequality and (4.4.14) and (4.4.15) we have

$$E_0|\xi_{\gamma_\rho}^0| \le 3E_0 \left[\int_0^{\gamma_\rho} |\pi_t|^2 \, dt\right]^{1/2}$$

$$= E_0 \left[\int_0^{\gamma_\rho} h^2(x_t) \sum_k |(\xi_t, \sigma^k(x_t))|^2 \, dt\right]^{1/2}$$

$$\le N E_0 \left[\int_0^{\gamma_\rho} h_1^2(x_t)|\xi_t|^2 \, dt\right]^{1/2}$$

$$\le N E_0 \sup_{t \le \gamma_\rho} |\xi_t|\psi_t^{1/2} \left[\int_0^{\gamma_\rho} h_1^2(x_s)\psi_s^{-1} \, ds\right]^{1/2}$$

$$\le N \left[E_0 \sup_{t \le \gamma_\rho} |\xi_t|^2\psi_t\right]^{1/2} \left[E_0 \int_0^{\gamma_\rho} h_1^2(x_s)\psi_s^{-1} \, ds\right]^{1/2} \le N|\xi|.$$

Finally, for any $\rho \in (0, 1)$

$$v_{(\xi)}(0) = E_0[v_{(\xi_{\gamma_\rho})}(x_{\gamma_\rho}) + \xi_{\gamma_\rho}^0 v(x_{\gamma_\rho})] \le |v_x|_{0,B}(1-\rho^2)|\xi| + N|v|_{0,B}|\xi|.$$

By letting $\rho \uparrow 1$ and using the equality $|v|_{0,B} = |g|_{0,\partial B}$ we obtain $v_{(\xi)}(0) \le N|g|_{0,\partial B}|\xi|$, which along with the arbitrariness of $\xi$ bring the proof to an end. $\qquad\square$

### 4.4.3. Parabolic equations

As in the case of Theorem 4.1.3 notationally it is convenient to treat parabolic equations as elliptic ones. The result below is known in the theory of PDE for continuous rather than smooth coefficients. In this subsection generic points in $\mathbb{R}^d$ with $d \ge 0$ are denoted by

$$x = (x', x^d), \quad x' \in \mathbb{R}^{d-1}, x^d \in \mathbb{R}.$$

THEOREM 4.4.5. *Let* $d \ge 2$,

$$D = \{x : x^d > 0\}, \quad , b^d \equiv -1, \quad \sigma^{dk} \equiv 0 \quad \text{for} \quad , i = 1, \ldots, d_1,$$
$$(a(x)\xi, \xi) \ge \varepsilon|\xi'|^2, \quad \forall x \in D, \xi \in \mathbb{R}^d,$$

*where* $\varepsilon > 0$ *is a constant. Then*

(4.4.16) $\qquad |v_{(\xi)}(x)| \le N\big((1 + (x^d)^{-\frac{1}{2}})|\xi'| + (1 + (x^d)^{-1})|\xi^d|\big) \sup_{x^d=0} |g|$

*in* $D$ *for any* $\xi \in \mathbb{R}^d$ *with a constant* $N$ *depending only on* $\varepsilon$, $d$, $d_1$ *and bounds for* $\sigma$, $b$ *and their first derivatives.*

PROOF. It suffices to concentrate on

(4.4.17) $$\xi = (\xi', 0).$$

This follows easily from Theorem 4.1.3 which implies that for $x^d \in (0, 2)$

$$|v_{x^d}(x)| \leq N(1 + (x^d)^{-\frac{1}{2}}) \sup_{y^d = x^d/2} |v_{y'}(y)|$$

and if $x^d \geq 2$, then

$$|v_{x^d}(x)| \leq N \sup_{y^d = x^d - 1} |v_{y'}(y)|.$$

In the rest of the proof we assume that (4.4.17) holds and introduce $\xi_t$ by using (4.4.10) with $\pi$ taken from (4.4.11). Since

$$x_t^d = x_0 - t,$$

we have $\xi_t^d = 0$ and this allows us to repeat major part of the proof of Theorem 4.1.3. By substituting there the origin with any $x_0 \in D$ without any trouble we come to (4.4.13) and (4.4.14) for any stopping time $\gamma \leq \tau = x_0^d$ if $h_1 \geq 1$.

After that we set

$$h_1(x) = 1 + \frac{1}{x^d}, \quad u(x) = 1 + \frac{1}{x^d}.$$

Then

$$a^{ij} u_{x^i x^j} + b^i u_{x^i} - 2h_1 u = -u_{x^d} - 2h_1 u \leq -h_1^2 \quad \text{in} \quad D$$

and instead of (4.4.15) for $\rho \in (0, x_0^d)$ and $\gamma_\rho$ defined as the first exit time from $\{x^d > \rho\}$ we obtain

$$u(x_0) \geq \left(1 + \frac{1}{\rho}\right) E_{x_0} \psi_{\gamma_\rho}^{-1} + E_{x_0} \int_0^{\gamma_\rho} h_1^2(x_t) \psi_t^{-1} \, dt.$$

Then as after (4.4.15)

$$(E_{x_0} |\xi_{\gamma_\rho}|)^2 \leq E_{x_0} |\xi_{\gamma_\rho}|^2 \psi_{\gamma_\rho} E_0 \psi_{\gamma_\rho}^{-1} \leq \frac{\rho}{1 + \rho} |\xi|^2 u(x_0),$$

$$E_{x_0} |\xi_{\gamma_\rho}^0| \leq N u^{1/2}(x_0) |\xi|,$$

$$v_{(\xi)}(x_0) = E_{x_0}[v_{(\xi_{\gamma_\rho})}(x_{\gamma_\rho}) + \xi_{\gamma_\rho}^0 v(x_{\gamma_\rho})]$$

$$\leq \sup_{x^d = \rho} |v_x| \rho |\xi| + N |v|_{0, D} u^{1/2}(x_0) |\xi|.$$

By letting $\rho \downarrow 0$ and using the equality $|v|_{0, D} = |g|_{0, \partial D}$ we obtain

$$v_{(\xi)}(0) \leq N u^{1/2}(x_0) |g|_{0, \partial D} |\xi|,$$

which along with the arbitrariness of $\xi$ proves (4.4.16) for $\xi = (\xi, 0)$. The theorem is proved.                                                                   $\square$

REMARK 4.4.6. Estimate (4.4.16) is sharp in what concerns the rate with which $v_{x'}(x)$ and $v_{x^d}(x)$ can blow up as $x^d \downarrow 0$.

To show that consider a diffusion corresponding to the heat equation. Let $d = 2$ and $d_1 = 1$. Define $x_t(x)$ as a solution of the "equation"

$$x_t^1 = x^1 + w_t, \quad x_t^2 = x^2 - t.$$

Then

$$v(x) = Eg(x^1 + w_{x^2}) = \int_{-\infty}^{\infty} g(y)p(y - x^1, x^2)\, dy,$$

where

$$p(x) = \frac{1}{\sqrt{2\pi x^2}} e^{-\frac{1}{2x^2}(x^1)^2}.$$

Hence

$$v_{x^1}(x) = \int_{-\infty}^{\infty} g(y)p_{x^1}(y - x^1, x^2)\, dy, \quad v_{x^2}(x) = \int_{-\infty}^{\infty} g(y)p_{x^2}(y - x^1, x^2)\, dy$$

and the best estimate of these quantities in terms of $\sup |g|$ are

$$|v_{x^1}(x)| \le \sup |g| \int_{-\infty}^{\infty} |p_{x^1}(y, x^2)|\, dy = (x^2)^{-1/2} \sup |g| \int_{-\infty}^{\infty} |p_{x^1}(y, 1)|\, dy,$$

$$|v_{x^2}(x)| \le \sup |g| \int_{-\infty}^{\infty} |p_{x^2}(y, x^2)|\, dy = (x^2)^{-1} \sup |g| \int_{-\infty}^{\infty} |p_{x^2}(y, 1)|\, dy,$$

where the equalities are obtained after changing variables $y \to yx^2$.

### 4.5. Some limitations of measure-change related quasiderivatives

Consider equation (3.0.1) without drift:

(4.5.1)                                $dx_t = \sigma^k(x_t)\, dw_t^k,$

take a smooth bounded function $g$, a $T \in (0, \infty)$, and let

$$v(s, x) = E_x g(x_{T-s}).$$

This is a particular case of processes from Theorem 4.4.5 which is seen if one adds an additional coordinate $s_t$ with the equation $ds_t = dt$ and introduces

$$D = \{(s, x) : s < T\}.$$

Here we discuss the possibility to estimate $|v_x|$ in terms of $\sup|g|$ by only using the measure-change related parameters $\pi$. Notice that the process

$$v(t, x_t)$$

is a martingale on $[0, T]$ so that for any quasiderivative $\xi_t$ of $x_t(x)$ on $[0, T)$ in the direction $\xi$ with adjoint $\xi_t^0$ the process

$$v_{(\xi_t)}(t, x_t) + \xi_t^0 v(t, x_t)$$

is a local martingale and under appropriate integrability assumption

(4.5.2) $$v_{(\xi)}(0, x) = E_x[g_{(\xi_T)}(x_T) + \xi_T^0 g(x_T)].$$

In Section 4.4 we have proved a well known result from the theory of PDE that there is a constant $N$ independent of $f$ such that

(4.5.3) $$\sup_x |v_x(0, x)| \leq N \sup_x |f(x)|$$

provided that $\sigma$ is uniformly nondegenerate. On the other hand, it is also well known that (4.5.3) holds if $\sigma^k$ satisfy the Hörmander condition which is much less restrictive than the uniform nondegeneracy although the smoothness assumptions are heavier. Therefore one could expect that under this condition one can find a quasiderivative $\xi_t$ such that

(4.5.4) $$\xi_T = 0.$$

We are going to show that most likely this is impossible if one only uses measure-change related quasiderivatives, which in our case are defined by the equation

$$d\xi_t = \sigma_{(\xi_t)}^k(x_t)\, dw_t^k - \sigma^k(x_t)\pi_t^k\, dt.$$

Since

$$d|\xi_t|^2 = 2(\xi_t, \sigma_{(\xi_t)}^k(x_t))\, dw_t^k + \left[\sum_k |\sigma_{(\xi_t)}^k(x_t)|^2 - 2(\xi_t, \sigma^k(x_t))\pi_t^k\right] dt$$

and we want to make $|\xi_T|$ as small as possible, it is natural to take

$$\pi_t^k = n(\xi_t, \sigma^k(x_t)),$$

where $n \geq 0$ is a number, so that

(4.5.5) $$d\xi_t = \sigma_{(\xi_t)}^k(x_t)\, dw_t^k - 2na(x_t)\xi_t dt,$$

(4.5.6) $$d|\xi_t|^2 = 2(\xi_t, \sigma_{(\xi_t)}^k(x_t))\, dw_t^k + \left[\sum_k |\sigma_{(\xi_t)}^k(x_t)|^2 - 4n(a(x_t)\xi_t, \xi_t)\right] dt,$$

$$|\xi_t|^2 + 4n \int_0^t (a(x_s)\xi_s, \xi_s)\, ds = |\xi|^2$$

(4.5.7)

$$+ 2 \int_0^t (\xi_s, \sigma^k_{(\xi_s)}(x_s))\, dw^k_s + \int_0^t \sum_k |\sigma^k_{(\xi_s)}(x_s)|^2\, ds\,.$$

Observe that by using Itô's formula to find the differential of $|\xi_t|^4 = (|\xi_t|^2)^2$ we immediately get that, for a constant $N$ independent of $t$ and $n$,

$$E|\xi_t|^4 \exp(-Nt)$$

is a decreasing function of $t$, so that

(4.5.8)                          $E|\xi_t|^4 \le |\xi|^4 e^{Nt} \quad \forall t, n \ge 0\,.$

One could expect that, as $n \to \infty$, having the term $(a(x_t)\xi_t, \xi_t)$ in (4.5.6) with large coefficient $n$ will yield that

(4.5.9)                          $|\xi_T| \xrightarrow{P} 0 \quad \text{as} \quad n \to \infty\,.$

Of course, one would want (4.5.9) to hold for any $T > 0$.

Observe, that (4.5.9) is certainly true when $a$ is uniformly nondegenerate, since then for an $\varepsilon > 0$ and all large $n$

$$(a(x_t)\xi_t, \xi_t) \ge \varepsilon|\xi_t|^2\,,$$

$$E|\xi_t|^2 \le |\xi|^2 - n\varepsilon \int_0^t E|\xi_s|^2\, ds, \quad E|\xi_t|^2 \le |\xi|^2 e^{-n\varepsilon t}\,.$$

If only the Hörmander condition is satisfied, in principle, it could possibly happen that still the boundedness of

(4.5.10)                          $n \int_0^T (a(x_t)\xi_t, \xi_t)\, dt\,,$

which follows from (4.5.7), will lead to (4.5.9). Actually, the boundedness of (4.5.10) if we replace $\xi_t$ with *constant* $\xi \ne 0$ is impossible if we *assume* that $x_T$ has density, which is the case under the Hörmander condition. Indeed, if we assume that with nonzero probability

$$\int_0^T (a(x_t)\xi, \xi)\, dt = 0\,,$$

then $\sigma^k(x_t) \perp \xi$ for almost all $t \in [0, T]$ and $x_T \perp \xi$ with positive probability.

To show that, generally, the above expectations are not realistic we first reduce equation (4.5.5) to the following "deterministic" one:

(4.5.11)                          $d\hat{\xi}_t = -2na(x_t)\hat{\xi}_t\, dt, \quad t \ge 0, \quad \hat{\xi}_0 = \xi\,.$

LEMMA 4.5.1. *Assume that (4.5.9) holds for any $T \in (0, \infty)$. Then*

$$|\hat{\xi}_T| \xrightarrow{P} 0 \quad \text{as} \quad n \to \infty$$

*for any $T \in (0, \infty)$.*

PROOF. The difference

$$\eta_t = \xi_t - \hat{\xi}_t$$

satisfies

$$d\eta_t = \sigma_{(\xi_t)}^k(x_t)\, dw_t^k - 2na(x_t)\eta_t\, dt, \quad t \geq 0, \quad \eta_0 = 0,$$

$$E|\eta_t|^2 \leq \sum_k E \int_0^t |\sigma_{(\xi_s)}^k(x_s)|^2\, ds \leq NE \int_0^t |\xi_s|^2\, ds.$$

The right-hand side here tends to zero due to the assumption of the lemma and (4.5.8), the latter implying the uniform integrability of $|\xi_s|^2$. Obviously this proves the lemma. $\square$

Equation (4.5.11) can be considered at each $\omega$. We will present an example in which for each $\xi \neq 0$ and $n > 0$, $|\hat{\xi}_t|$ is a strictly decreasing function, so that equation (4.5.11) possesses certain dumping property, however, for almost all initial data, $|\hat{\xi}_t| \not\to 0$ as $n \to \infty$. This example came as a surprise to the author and few specialists in dynamical systems he contacted.

Consider the diffusion process corresponding to the so-called Heisenberg-Laplacian in

$$\mathbb{R}^3 = \{(x, y, z) : x, y, z \in \mathbb{R}\}.$$

It is a well-known and easy fact that this operator satisfies the Hörmander condition. The reader noticed that we changed notation. This is because the coordinates play here quite different roles.

Our process is given by the equation

$$d \begin{pmatrix} x_t \\ y_t \\ z_t \end{pmatrix} = \begin{pmatrix} 1 \\ 0 \\ -y_t \end{pmatrix} dw_t + \begin{pmatrix} 0 \\ 1 \\ x_t \end{pmatrix} dB_t,$$

with some nonrandom initial data $(x_0, y_0, z_0) \in \mathbb{R}^3$, where $B_t$ and $w_t$ are independent one-dimensional Wiener processes. Here

$$2a(x, y, z) = \sigma(x, y, z)\sigma^*(x, y, z) = \begin{pmatrix} 1 & 0 & -y \\ 0 & 1 & x \\ -y & x & x^2 + y^2 \end{pmatrix}$$

and equation (4.5.11) becomes

(4.5.12)
$$d \begin{pmatrix} \xi_t \\ \eta_t \\ \zeta_t \end{pmatrix} = -n \begin{pmatrix} \xi_t & & -\zeta_t y_t \\ & \eta_t & +\zeta_t x_t \\ -\xi_t y_t & +\eta_t x_t & +\zeta_t(x_t^2 + y_t^2) \end{pmatrix} dt.$$

THEOREM 4.5.2.
(i) If $\xi_0^2 + \eta_0^2 + \zeta_0^2 \neq 0$, then for any $n > 0$ and $t > s \geq 0$ (a.s.)

(4.5.13)
$$\xi_t^2 + \eta_t^2 + \zeta_t^2 < \xi_s^2 + \eta_s^2 + \zeta_s^2.$$

(ii) *As* $n \to \infty$, *uniformly on* $[0, 1]$ *in probability*

$$\xi_t - e^{-nt}(\xi_0 - y_0\bar{\zeta}_0 - y_0\psi_t) \to \bar{\zeta}_t y_t,$$
$$\eta_t - e^{-nt}(\eta_0 + x_0\bar{\zeta}_0 + x_0\psi_t) \to -\bar{\zeta}_t x_t,$$
$$\zeta_t + (\bar{\zeta}_0 - \zeta_0)\phi_t \to \bar{\zeta}_t,$$

*where*

$$\psi_t = (\bar{\zeta}_0 - \zeta_0)n \int_0^t \exp\left(-n \int_0^s r_u^2 \, du\right) ds, \quad \phi_t = \exp\left(-n \int_0^t (1 + r_s^2) \, ds\right),$$

$$r_t^2 = x_t^2 + y_t^2, \qquad\qquad\qquad \bar{\zeta}_0 = \frac{\xi_0 y_0 - \eta_0 x_0 + \zeta_0}{1 + r_0^2},$$

$$\bar{\zeta}_t := \bar{\zeta}_0 \frac{\sqrt{1 + r_0^2}}{\sqrt{1 + r_t^2}} \exp\left(-(1/2) \int_0^t \frac{2 + r_s^2}{(1 + r_s^2)^2} \, ds\right).$$

*In particular, for any* $t > 0$

$$\xi_t^2 + \eta_t^2 + \zeta_t^2 \to \bar{\zeta}_t^2(1 + r_t^2) = \frac{(\xi_0 y_0 - \eta_0 x_0 + \zeta_0)^2}{1 + r_0^2} \exp\left(-\int_0^t \frac{2 + r_s^2}{(1 + r_s^2)^2} \, ds\right).$$

We need some work to be done before proving the theorem. First, observe that the right-hand side of (4.5.12) can be expressed in terms of the processes

$$\mu_t := \xi_t - \zeta_t y_t, \quad \nu_t := \eta_t + \zeta_t x_t.$$

By taking into account that

$$-\xi_t y_t + \eta_t x_t + \zeta_t(x_t^2 + y_t^2) = -y_t(\xi_t - \zeta_t y_t) + x_t(\eta_t + \zeta_t x_t)$$

we rewrite (4.5.12) as

(4.5.14)      $d\xi_t = -n\mu_t \, dt, \quad d\eta_t = -n\nu_t \, dt, \quad d\zeta_t = n(y_t\mu_t - x_t\nu_t) \, dt.$

Next we note a result of straightforward computations.

LEMMA 4.5.3. *We have*

(4.5.15)      $\xi_t^2 + \eta_t^2 + \zeta_t^2 = \xi_0^2 + \eta_0^2 + \zeta_0^2 - 2n \int_0^t (\mu_s^2 + \nu_s^2) \, ds.$

*In particular,* $\xi_s^2 + \eta_s^2 + \zeta_s^2$ *is bounded and*

$$\alpha_t := x_t\mu_t + y_t\nu_t$$

*tends to zero in* $L_2([0, T])$ *for any* $\omega$ *and* $T$ *as* $n \to \infty$.

PROOF OF THEOREM 4.5.2 (i). The uniqueness of solutions of linear ordinary equations implies that

$$\xi_s^2 + \eta_s^2 + \zeta_s^2 \neq 0 .$$

Next, if on a set $A$ of positive probability we have an equality, then owing to (4.5.15)

$$\xi_r - \zeta_r y_r = \eta_r + \zeta_r x_r = 0 \quad s \leq r \leq t \quad \text{on} \quad A .$$

By using this in (4.5.12) we find that

$$\xi_r = \xi_s, \quad \eta_r = \eta_s, \quad \zeta_r = \zeta_s$$

for $s \leq r \leq t$ on $A$. But now the third equation in (4.5.12) shows that

$$\xi_s y_r + \eta_s x_r + \zeta_s (x_r^2 + y_r^2) = 0$$

for $s \leq r \leq t$ with positive probability. By remembering that

$$x_t = x_0 + w_t, \quad y_t = y_0 + B_t$$

we easily see that this can only happen if $\xi_s^2 + \eta_s^2 + \zeta_s^2 = 0$ and assertion (i) is proved.

The proof of assertion (ii) of Theorem 4.5.2 is much longer. First write the system in terms of $\mu_t$ and $v_t$. We have

$$\begin{aligned}
d\mu_t &= d\xi_t - y_t\, d\zeta_t - \zeta_t\, dy_t = -\zeta_t\, dy_t - n\mu_t\, dt - y_t n(y_t\mu_t - x_t v_t)\, dt \\
&= -\zeta_t\, dy_t - n[(1 + y_t^2)\mu_t - y_t x_t v_t]\, dt \\
&= -\zeta_t\, dy_t - n[(1 + r_t^2)\mu_t - x_t \alpha_t]\, dt , \\
dv_t &= \zeta_t\, dx_t - nv_t\, dt + nx_t(y_t\mu_t - x_t v_t)\, dt \\
&= \zeta_t\, dx_t - n[(1 + x_t^2)v_t - y_t x_t \mu_t]\, dt \\
&= \zeta_t\, dx_t - n[(1 + r_t^2)v_t - y_t \alpha_t]\, dt .
\end{aligned}$$

In short,

$$\begin{aligned}
d\mu_t &= -\zeta_t\, dy_t - n[(1 + r_t^2)\mu_t - x_t \alpha_t]\, dt , \\
dv_t &= \zeta_t\, dx_t - n[(1 + r_t^2)v_t - y_t \alpha_t]\, dt , \\
d\zeta_t &= n(y_t\mu_t - x_t v_t)\, dt .
\end{aligned}$$

Two first equations yield

$$\mu_t = (\xi_0 - \zeta_0 y_0)\phi_t - \phi_t \int_0^t \phi_s^{-1}\zeta_s\, dy_s + n \int_0^t \phi_t\phi_s^{-1}x_s\alpha_s\, ds ,$$

$$v_t = (\eta_0 + \zeta_0 x_0)\phi_t + \phi_t \int_0^t \phi_s^{-1}\zeta_s\, dx_s + n \int_0^t \phi_t\phi_s^{-1}y_s\alpha_s\, ds ,$$

where

$$\phi_t = \exp\left(-n \int_0^t (1 + r_s^2)\, ds\right) .$$

We see that

$$\zeta_t = \zeta_0 + P(t) + Q(t)$$

(4.5.16)
$$- n \int_0^t y_s \phi_s \left( \int_0^s \phi_r^{-1} \zeta_r \, dy_r \right) ds$$

$$- n \int_0^t x_s \phi_s \left( \int_0^s \phi_r^{-1} \zeta_r \, dx_r \right) ds \, ,$$

where

$$P(t) := n \int_0^t \left[ n y_s \int_0^s \phi_s \phi_r^{-1} x_r \alpha_r \, dr - n x_s \int_0^s \phi_s \phi_r^{-1} y_r \alpha_r \, dr \right] ds \, ,$$

$$Q(t) := (\xi_0 - \zeta_0 y_0) n \int_0^t \phi_s y_s \, ds - (\eta_0 + \zeta_0 x_0) n \int_0^t \phi_s x_s \, ds \, .$$

First we deal with $P(t)$. Integrate in $s$ to find that

$$n^2 \int_0^t y_s \int_0^s \phi_s \phi_r^{-1} x_r \alpha_r \, dr ds$$

$$= n \int_0^t \phi_r^{-1} x_r \alpha_r \int_r^t \frac{y_s}{1 + r_s^2} \, d \exp\left( -n \int_0^s (1 + r_u^2) \, du \right) dr$$

$$= -n \int_0^t \phi_r^{-1} x_r \alpha_r \left( \int_r^t \phi_s \, d \frac{y_s}{1 + r_s^2} \right) dr$$

$$+ n \frac{y_t}{1 + r_t^2} \phi_t \int_0^t \phi_r^{-1} x_r \alpha_r \, dr - n \int_0^t \frac{x_r y_r}{1 + r_r^2} \alpha_r \, dr =: I_1(t) + I_2(t) + I_3(t) \, .$$

Similarly,

$$n^2 \int_0^t x_s \int_0^s \phi_s \phi_r^{-1} y_r \alpha_r \, dr ds$$

$$= n \int_0^t \phi_r^{-1} y_r \alpha_r \int_r^t \frac{x_s}{1 + r_s^2} \, d \exp\left( -n \int_0^s (1 + r_u^2) \, du \right) dr$$

$$= -n \int_0^t \phi_r^{-1} y_r \alpha_r \left( \int_r^t \phi_s d \frac{x_s}{1 + r_s^2} \right) dr$$

$$+ n \frac{x_t}{1 + r_t^2} \phi_t \int_0^t \phi_r^{-1} y_r \alpha_r \, dr - n \int_0^t \frac{x_r y_r}{1 + r_r^2} \alpha_r \, dr =: J_1(t) + J_2(t) + I_3(t) \, .$$

Now we can prove the following.

LEMMA 4.5.4. *We have*

(i)
$$I_1(t) = - \int_0^t \left[ n \int_0^s \phi_s \phi_r^{-1} x_r \alpha_r \, dr \right] d \frac{y_s}{1 + r_s^2} \, ,$$

(ii) $\sup_{t \le T} [|I_1(t)| + |J_1(t)|] \to 0$ *in probability*,
(iii) $\sup_{t \le T} |I_2(t) - J_2(t)| \to 0$ *in probability*,
(iv) $\sup_{t \le T} |P(t)| \to 0$ *in probability*.

PROOF. Assertion (i) is obvious. Assertion (ii) follows from Lemma 4.5.3 and the fact that

$$\left| n \int_0^s \phi_s \phi_r^{-1} x_r \alpha_r \, dr \right| \leq n \int_0^s e^{-n(s-r)} x_r \alpha_r \, dr$$

where the last expression is a convolution with an $L_1$-norm one function.

Assertion (iii) is proved by the following estimates which are also available for $J_2$

$$\left| I_2(t) - n \frac{x_t y_t}{1+r_t^2} \int_0^t \phi_t \phi_r^{-1} \alpha_r \, dr \right| \leq \left| n \int_0^t \phi_t \phi_r^{-1} [x_r - x_t] \alpha_r \, dr \right|$$

$$\leq n \int_0^t e^{-n(t-r)} |x_r - x_t| \, dr \sup |\alpha| \to 0.$$

Assertion (iv) is an obvious consequence of the previous ones. The lemma is proved. □

LEMMA 4.5.5. *The processes*

$$Q(t) + \frac{\xi_0 y_0 - \eta_0 x_0 - \zeta_0 r_0^2}{1+r_0^2} \phi_t = Q(t) + (\bar{\zeta}_0 - \zeta_0) \phi_t$$

*are uniformly bounded and tend as $n \to \infty$ uniformly on $[0, 1]$ in probability to their common value at zero: $\bar{\zeta}_0 - \zeta_0$.*

PROOF. Observe that, for instance,

$$n \int_0^t \phi_s y_s \, ds = - \int_0^t \frac{y_s}{1+r_s^2} \, d\phi_s = \frac{y_0}{1+r_0^2} - \frac{y_t}{1+r_t^2} \phi_t + \int_0^t \phi_s \, d \frac{y_s}{1+r_s^2}.$$

Here the last term tends to zero in probability since $\phi_s \to 0$ for $s \neq 0$. Also

$$\frac{y_t}{1+r_t^2} \phi_t = \frac{y_0}{1+r_0^2} \phi_t + \left[ \frac{y_t}{1+r_t^2} - \frac{y_0}{1+r_0^2} \right] \phi_t,$$

where the second term on the right tends to zero uniformly due to the continuity of $x_t$ and $y_t$. Treating the other integral in the definition of $Q(t)$ in the same way we easily get our assertion. The lemma is proved. □

To deal with the remaining terms in (4.5.16) observe that

$$n \int_0^t y_s \phi_s \left( \int_0^s \phi_r^{-1} \zeta_r \, dy_r \right) ds = - \int_0^t \frac{y_s}{1+r_s^2} \left( \int_0^s \phi_r^{-1} \zeta_r \, dy_r \right) d\phi_s$$

$$= - \frac{y_t}{1+r_t^2} \beta_t + \int_0^t \zeta_s \frac{y_s}{1+r_s^2} \, dy_s$$

$$+ \int_0^t \beta_s \, d \frac{y_s}{1+r_s^2} + \int_0^t \zeta_s \left( d \frac{y_s}{1+r_s^2} \right) dy_s,$$

where

$$\beta_t = \phi_t \int_0^t \phi_r^{-1} \zeta_r \, dy_r.$$

LEMMA 4.5.6. *We have $\sup_{t \leq T} |\beta_t| \to 0$ in probability.*

This lemma obviously follows from Theorem 4.5.7 below. Also notice that

$$\left(d\frac{y_s}{1+r_s^2}\right)dy_s = \left[\frac{1}{1+r_s^2} - \frac{2y_s^2}{(1+r_s^2)^2}\right]ds = \frac{1+x_s^2-y_s^2}{(1+r_s^2)^2}\,ds\,.$$

We now see that, for a certain process $\gamma_t$, which goes to zero uniformly on $[0,1]$ in probability, we have

$$\beta_t := \zeta_t + (\bar\zeta_0 - \zeta_0)\phi_t$$
$$= \gamma_t + \bar\zeta_0 - \int_0^t \zeta_s\frac{y_s\,dy_s + x_s\,dx_s}{1+r_s^2} - 2\int_0^t \zeta_s\frac{ds}{(1+r_s^2)^2}\,.$$

Now it is a trivial matter to show that $\beta_t$ uniformly on $[0,1]$ converges in probability to the unique solution $\bar\beta_t$ of

$$\bar\beta_t = \bar\zeta_0 - \int_0^t \bar\beta_s\frac{y_s\,dy_s + x_s\,dx_s}{1+r_s^2} - 2\int_0^t \bar\beta_s\frac{ds}{(1+r_s^2)^2}\,.$$

The solution is written as

$$\bar\beta_t = \bar\zeta_0\exp\left(-\int_0^t \frac{1}{1+r_s^2}(y_s\,dy_s + x_s\,dx_s) - (1/2)\int_0^t \frac{r_s^2+4}{(1+r_s^2)^2}\,ds\right)\,.$$

On the other hand, Itô's formula shows that

$$d\ln(1+r_t^2) = \frac{dr_t^2}{1+r_t^2} - \frac{1}{2(1+r_t^2)^2}(dr_t^2)^2\,,$$
$$dr_t^2 = 2x_t\,dx_t + 2y_t\,dy_t + 2\,dt\,,\qquad (dr_t^2)^2 = 4r_t^2\,dt\,,$$

which implies that

$$\frac{1}{2}d\ln(1+r_t^2) = \frac{1}{1+r_t^2}(y_t\,dy_t + x_t\,dx_t) + \left(\frac{1}{1+r_t^2} - \frac{r_t^2}{(1+r_t^2)^2}\right)dt$$
$$= \frac{1}{1+r_t^2}(y_t\,dy_t + x_t\,dx_t) + \frac{1}{(1+r_t^2)^2}\,dt\,,$$
$$- \int_0^t \frac{1}{1+r_s^2}(y_s\,dy_s + x_s\,dx_s)$$
$$= \int_0^t \frac{1}{(1+r_s^2)^2}\,ds + \ln\sqrt{1+r_0^2} - \ln\sqrt{1+r_t^2}\,.$$

It follows that

$$\bar\beta_t = \bar\zeta_0\frac{\sqrt{1+r_0^2}}{\sqrt{1+r_t^2}}\exp\left(-(1/2)\int_0^t \frac{r_s^2+2}{(1+r_s^2)^2}\,ds\right) = \bar\zeta_t\,.$$

This proves the part of Theorem 4.5.2 (ii) concerning $\zeta_t$.

Next, observe that

$$\xi_t = \xi_0 e^{-nt} + n \int_0^t e^{-n(t-s)} \zeta_s y_s \, ds \, ,$$

$$n \int_0^t e^{-n(t-s)} \zeta_s y_s \, ds = n \int_0^t e^{-n(t-s)} \beta_s y_s \, ds$$

$$- y_0 e^{-nt} \psi_t - (\bar{\zeta}_0 - \zeta_0) n \int_0^t e^{-n(t-s)} (y_s - y_0) \phi_s \, ds \, .$$

The last term tends to zero since $(y_s - y_0)\phi_s \to 0$ uniformly on $[0, 1]$. Furthermore,

$$n \int_0^t e^{-n(t-s)} \beta_s y_s \, ds - \bar{\zeta}_t y_t = n \int_0^t e^{-n(t-s)} [\beta_s y_s - \bar{\zeta}_t y_t] \, ds - \bar{\zeta}_t y_t e^{-nt} \, .$$

Here the first term on the right goes to zero uniformly in $t$ because $\beta_s y_s \to \bar{\zeta}_s y_s$ uniformly. As for the second term, notice that $[\bar{\zeta}_t y_t - \bar{\zeta}_0 y_0]e^{-nt} \to 0$ uniformly. This takes care of the process $\xi_t$.

Similarly we have

$$\eta_t = \eta_0 e^{-nt} - n \int_0^t e^{-n(t-s)} \zeta_s x_s \, ds \, ,$$

and just by replacing above $y_t$ with $-x_t$ we get the result for $\eta_t$.

Theorem 4.5.2 is proved.

THEOREM 4.5.7. *Let $\kappa \in (0, 1/2)$, $p > 2$, and $v \geq 1$ be some constants, $\gamma_t$ an Itô integrable process satisfying $E|\gamma_t|^p \leq \mu^p$, where $\mu^p$ is a constant. Let $v_t$ be an adapted locally integrable process such that $v_t \geq v$ and introduce*

$$\varphi_t = \exp\left(-\int_0^t v_s \, ds\right) \, .$$

*Then there is a constant $N$, depending only on $p$ and $\kappa$, such that*

$$E \sup_{t \leq 1} \left| \varphi_t \int_0^t \varphi_s^{-1} \gamma_s \, dB_s \right|^p \leq N \mu^p v^{-(p-2)\kappa} \, .$$

PROOF. By integrating by parts we see that

$$\varphi_t \int_0^t \varphi_s^{-1} \gamma_s \, dB_s = m_t \varphi_t + \int_0^t \varphi_t \varphi_s^{-1} [m_t - m_s] \, ds =: I_t + J_t \, ,$$

where

$$m_t := \int_0^t \gamma_t \, dB_t \, .$$

To estimate $J_t$ introduce $\delta = \kappa(p-2)/p$ and observe that

$$|J_t| \leq \int_0^t e^{-\nu(t-s)}|m_t - m_s|\,ds \leq R \int_0^t e^{-\nu(t-s)}(t-s)^\delta\,ds$$

$$\leq R \int_0^\infty e^{-\nu s}s^\delta\,ds = NR\nu^{-(\delta+1)} \leq NR\nu^{-(p-2)\kappa/p},$$

where

$$R := \sup_{0 \leq u < v \leq 1} \frac{1}{(v-u)^\delta}|m_v - m_u|.$$

Also, obviously for $t \leq 1$

$$|I_t| \leq R \sup_{t \leq 1} t^\delta e^{-\nu t} \leq R\nu^{-\delta} \sup_{t \geq 0} t^\delta e^{-t}.$$

Hence, to prove the lemma it only remains to show that

$$ER^p \leq N\mu^p.$$

By embedding theorems (see, for instance, Lemma II.2.4 in [12]) for $q \geq 1$, $\chi > 1/q$, and $u, v \in [0, 1]$ we have

$$\frac{|m_v - m_u|^q}{(v-u)^{q\chi-1}} \leq N \int_0^1 \int_0^1 \frac{|m_t - m_s|^q}{|t-s|^{1+q\chi}}\,dt\,ds,$$

where the constant $N$ depends only on $q$ and $\chi$. For $q = p$ and $p\chi - 1 = p\delta$ we have $\chi = 1/p + \delta > 1/q$, so that

(4.5.17)                $$ER^p \leq N \int_0^1 \int_0^1 \frac{E|m_t - m_s|^p}{|t-s|^{2+p\delta}}\,dt\,ds.$$

Next, notice that since $E|\gamma_t|^p \leq \mu^p$, by trivial estimates of moments of stochastic integrals

$$E \sup_{s \leq r \leq t} |m_r - m_s|^p \leq N(t-s)^{p/2}\mu^p.$$

This shows that the right-hand side of (4.5.17) is less than

$$N\mu^p \int_0^1 \int_0^1 \frac{1}{|t-s|^{2+p\delta-p/2}}\,dt\,ds < \infty,$$

where the last inequality holds because $2 + p\delta - p/2 < 1$. The theorem is proved.                                                                    □

## 5. Epilog: new ways of using quasiderivatives

The goal of this chapter is to develop universal probabilistic methods of dealing with problems from Prolog, Subsection 4.2.1, and Section 4.3 even if the corresponding operator contains constant drift term. One may also hope that these methods will allow one to give a new probabilistic proof of the Hörmander theorem on hypoellipticity. Here we closely follow [14].

### 5.1.  A general result

In this section $D$ is a bounded $C^1$ domain in $\mathbb{R}^d$. We assume that $\sigma$ and $b$ satisfy Lipschitz condition and in $D$

(5.1.1)
$$\|\sigma\|^2 + |b| \leq K ,$$
$$E_x \tau \leq K, \quad E_x \tau \leq K \operatorname{dist}(x, \partial D),$$

where $K$ is a constant.

First we give a probabilistic counterpart of Lemma 1.0.5 for general processes.

LEMMA 5.1.1. *Let $p \in C^2(\bar{D})$, $g \in C^1(\bar{D})$. Introduce*

$$u(x) = E_x g(x_\tau), \quad v(x) = E_x(pg)(x_\tau)$$

*and assume that $u, v \in C^1(\bar{D})$. Then on $\partial D$ for any $\eta \in \mathbb{R}^d$ we have*

(5.1.2)          $$|v_{(\eta)} - (pu)_{(\eta)}| \leq N|g|'_{1,D}(|p_x|_{0,D} + |Lp|_{0,D})|(n, \eta)| ,$$

*where $n$ is the unit inward normal on $\partial D$ and $N$ depends only on $K$.*

PROOF. First observe the following standard computations using the Markov property: for any integer $n \geq 2$

$$E_x \tau^n = n E_x \int_0^\infty (\tau - t)^{n-1} I_{\tau > t} \, dt = n E_x \int_0^\infty I_{\tau > t} E_{x_t} \tau^{n-1} \, dt$$

$$\leq n \sup_{y \in D} E_y \tau^{n-1} E_x \int_0^\infty I_{\tau > t} \, dt = n E_x \tau \sup_{y \in D} E_y \tau^{n-1} ,$$

which by induction implies that

(5.1.3)          $$E_x \tau^n \leq n! \sup_{y \in D}(E_y \tau)^n, \quad E_x \tau^n \leq N E_x \tau \leq N \operatorname{dist}(x, \partial D) .$$

Next, notice that as in Lemma 1.0.5 it suffices to concentrate on $\eta = n$, fix a $y \in \partial D$, and choose $\varepsilon_0 > 0$ so that $y + \varepsilon n \in D$ as long as $0 < \varepsilon \leq \varepsilon_0$. Also fix an $\varepsilon \in (0, \varepsilon_0]$ and for $x_0 = x := y + \varepsilon n$ write

$$p(x_\tau) = p(x_0) + \int_0^\tau p_{(\sigma^k)}(x_s) \, dw_s^k + \int_0^\tau Lp(x_s) \, ds .$$

Furthermore,

$$I(x) := E_x g(x_\tau) \int_0^\tau P_{(\sigma^k)}(x_s) \, dw_s^k = E_x[g(x_\tau) - g(x)] \int_0^\tau P_{(\sigma^k)}(x_s) \, dw_s^k$$

$$\leq [g]_{1,D} \left( E_x |x_\tau - x|^2 \right)^{1/2} \left( E_x \int_0^\tau |\sigma^* p_x(x_s)|^2 \, ds \right)^{1/2} .$$

Here

$$E_x |x_\tau - x|^2 \leq 2E_x \left| \int_0^\tau \sigma(x_t) \, dw_t \right|^2 + 2K^2 E_x \tau^2 \leq N \operatorname{dist}(x, \partial D) ,$$

so that

$$I(x) \leq N[g]_{1,D} |\sigma^* p_x|_{0,D} \operatorname{dist}(x, \partial D) .$$

Now,

(5.1.4)   $v(x) = E_x(pg)(x_\tau) = p(x) E_x g(x_\tau) + I(x) + E_x g(x_\tau) \int_0^\tau Lp(x_s) \, ds .$

Here $E_x g(x_\tau) = u(x)$ and the last term on the right in (5.1.4) is less than

$$K|g|_{0,D} |Lp|_{0,D} \operatorname{dist}(x, \partial D) .$$

Hence we infer from (5.1.4) that

(5.1.5)   $v(y + \varepsilon n) \leq p u(y + \varepsilon n) + N([g]_{1,D} |p_x|_{0,D} + |g|_{0,D} |Lp|_{0,D}) \operatorname{dist}(x, \partial D) .$

Upon subtracting from this inequality the equality $v(y) = p(y) u(y)$, dividing through the result by $\varepsilon$, and letting $\varepsilon \downarrow 0$, we arrive at

$$v_{(n)}(y) - (pu)_{(n)}(y) \leq N([g]_{1,D} |p_x|_{0,D} + |g|_{0,D} |Lp|_{0,D}) .$$

Replacing $g$ with $-g$ yields an estimate of $v_{(n)} - (pu)_{(n)}$ from below, which being combined with the above result leads to (5.1.2) and proves the lemma.  □

COROLLARY 5.1.2. *If $g \equiv 1$, then $u \equiv 1$ so that $u_{(n)} = 0$ and*

$$|v_{(n)}| \leq N|p|_{2,D} .$$

REMARK 5.1.3. *For any $\beta \geq 1$ we have $\tau^\beta \leq \tau + \tau^n$, where $n = [\beta + 1]$. Hence (5.1.3) implies that*

$$E_x \tau^\beta \leq N \operatorname{dist}(x, \partial D) ,$$

*where $N$ depends only on $K$ and $\beta$.*

COROLLARY 5.1.4. *Let $\xi \in \mathbb{R}^d$, $x_0 \in D$. Under the assumptions of the lemma let $\xi_t$ be a quasiderivative of $x_t(x)$ at $x_0$ with adjoint process $\xi_t^0$ and $\xi_0 = \xi$. Assume that*

$$\xi_{t \wedge \tau(x_0)}, \quad \xi_{t \wedge \tau(x_0)}^0$$

*are uniformly integrable. Then*

(5.1.6)
$$|v_{(\xi)}(x_0) - E_{x_0}\xi_\tau^0 pg(x_\tau) - E_{x_0} pu_{(\xi_\tau)}(x_\tau)|$$
$$\leq N|g|'_{1,D}(|p_x|_{0,D}E_{x_0}|\xi_\tau| + |Lp|_{0,D}E_{x_0}|(n(x_\tau), \xi_\tau)|).$$

Indeed, by definition

$$v_{(\xi)}(x_0) = E_{x_0}v_{(\xi_\tau)}(x_\tau) + E_{x_0}\xi_\tau^0 v(x_\tau),$$

where

$$v(x_\tau) = pu(x_\tau) = pg(x_\tau) \quad \text{and} \quad v_{(\xi_\tau)}(x_\tau) = (pu)_{(\xi_\tau)}(x_\tau) + I$$

with

$$|I| \leq N|g|'_{1,D}(|p_x|_{0,D} + |Lp|_{0,D})|(n(x_\tau), \xi_\tau)|$$

and

$$|(pu)_{(\xi_\tau)}(x_\tau) - pu_{(\xi_\tau)}(x_\tau)| = |up_{(\xi_\tau)}(x_\tau)| \leq |g|_{0,D}|p_x|_{0,D}|\xi_\tau|.$$

THEOREM 5.1.5. *Let some functions*

$$p^{(1)}, \dots, p^{(m)} \in C^2(\bar{D}), \quad g, q \in C^1(\bar{D}),$$

*a vector $\xi \in \mathbb{R}^d$ and a point $x_0 \in D$. Assume that $q > 0$ in $\bar{D}$ and on $\partial D$ we have*

$$q(x) = \sum_{k=1}^{m} p^{(k)}(x).$$

*Introduce*

$$\bar{g} = g/q, \quad u(x) = E_x g(x_\tau), \quad \bar{u} = E_x \bar{g}(x_\tau), \quad v^{(k)}(x) = E_x p^{(k)}\bar{g}(x_\tau)$$

*and assume that*

$$u, \bar{u}, v^{(1)}, \dots, v^{(m)} \in C^1(\bar{D}).$$

*Let $\xi_t^{(k)}$, $k = 1, \dots, m$, be the first quasiderivatives of $x_t(x)$ at point $x_0$ with adjoint processes $\xi_t^{(0k)}$ and $\xi_0^{(k)} = \xi$. Assume that, for $\tau = \tau(x_0)$, the processes*

$$\xi_{t \wedge \tau}^{(k)}, \quad \xi_{t \wedge \tau}^{(0k)}$$

*are uniformly integrable and (a.s.)*

(5.1.7)
$$\bar{\xi}_\tau := \sum_{k=1}^{m} p^{(k)}(x_\tau)\xi_\tau^{(k)} \perp n(x_\tau).$$

*Then we have*

(5.1.8)
$$|u_{(\xi)}(x_0)| \leq \sum_{k=1}^{m} (|p^{(k)}\bar{g}|_{0,D}E_{x_0}|\xi_\tau^{(0k)}| + |p^{(k)}\bar{g}_x|_{0,D}E_{x_0}|\xi_\tau^{(k)}|)$$
$$+ N|\bar{g}|'_{1,D}\sum_{k=1}^{m} (|p_x^{(k)}|_{0,D} + |Lp^{(k)}|_{0,D})E_{x_0}|\xi_\tau^{(k)}|,$$

*where $N$ depends only on $K$.*

PROOF. By Corollary 5.1.4

$$|v_{(\xi)}^{(k)}(x_0) - E_{x_0}p^{(k)}\bar{g}(x_\tau)\xi_\tau^{(0k)} - E_{x_0}p^{(k)}\bar{u}_{(\xi_\tau^{(k)})}(x_\tau)|$$
$$\leq N|\bar{g}|'_{1,D}(|p_x^{(k)}|_{0,D} + |Lp^{(k)}|_{0,D})E_{x_0}|\xi_\tau^{(k)}|.$$

We sum up these inequalities with respect to $k$ and observe that

$$\sum_{k=1}^{m} v^{(k)}(x) = E_x\bar{g}\sum_{k=1}^{m}p^{(k)}(x_\tau) \equiv u(x)$$

and owing to (5.1.7)

$$\sum_{k=1}^{m} p^{(k)}(x_\tau)\bar{u}_{(\xi_\tau^{(k)})}(x_\tau) = \bar{u}_{(\bar{\xi}_\tau)}(x_\tau)$$

$$= \bar{g}_{(\bar{\xi}_\tau)}(x_\tau) = \sum_{k=1}^{m} p^{(k)}(x_\tau)\bar{g}_{(\xi_\tau^{(k)})}(x_\tau).$$

Then we immediately get (5.1.8) and the theorem is proved.　　□

REMARK 5.1.6. The most natural choice for $q$ is $q \equiv 1$. However in our applications this restriction leads to slightly worse results.

## 5.2. Equations without drift in strictly convex domains

Here we consider equation (0.0.1) with constant $\sigma$ assuming that

$$\text{tr}\,\sigma\sigma^* = 1, \quad b \equiv 0.$$

Let $D \in C^3$ be a uniformly convex domain in $\mathbb{R}^d$. Then there exists a concave function $\psi \in C^3(\bar{D})$ such that $\psi > 0$ in $D$, $\psi = 0$ on $\partial D$ and

$$\psi_{(l)(l)} \leq -2 \quad \text{in} \quad D$$

for any unit $l \in \mathbb{R}^d$. For such a function we have $L\psi \leq -1$, which along with Itô's formula imply that

$$E_x\tau \leq \psi(x),$$

so that assumption (5.1.1) is satisfied with a constant depending only on $|\psi|_{2,D}$. By the way, observe that the diameter of $D$ also can be easily estimated through $|\psi|_{2,D}$.

THEOREM 5.2.1. *Let* $g \in C^{0,1}(\bar{D})$. *Then* $u \in C^{0,1}_{\mathrm{loc}}(D)$ *and there is a constant* $N$ *depending only on* $|\psi|_{3,D}$ *and* $d$ *such that*

$$|u_x| \le N\psi^{-2}|g|_{1,D}.$$

PROOF. It suffices to prove the theorem for nondegenerate $a$ because as long as $N$ is independent of $a$, one can always pass to the limit approximating a degenerate $a$ with nondegenerate ones. For a similar reason we may assume that $g$ and $D$ are infinitely differentiable. In that case $u$ is infinitely differentiable and we fix an $x_0 \in D$ and $\xi \in \mathbb{R}^d$ with the goal in mind to estimate $u_{(\xi)}(x_0)$.

We are going to only use quasiderivatives based on time change associated with parameter $r$. Take a constant $r > 0$ and let

$$d\xi_t^{(1)} = r\sigma^k \, dw_t^k = r \, dx_t, \qquad d\xi_t^{(2)} = -r\sigma^k \, dw_t^k = -r \, dx_t,$$
$$\xi_t^{(1)} = \xi + r(x_t - x_0), \qquad \xi_t^{(2)} = \xi - r(x_t - x_0), \qquad \xi_t^{(0i)} \equiv 0.$$

Clearly, a vector $y \in \mathbb{R}^d$ is tangential to $\partial D$ at a point $x \in \partial D$ if and only if $\psi_{(y)}(x) = 0$. Therefore, to satisfy (5.1.7) with $m = 2$ we need to find two functions $p^{(1)}$ and $p^{(2)}$ in $\bar{D}$ such that, for $x \in \partial D$,

(5.2.1)
$$0 = p^{(1)}\psi_{(\xi + r(x - x_0))}(x) + p^{(2)}\psi_{(\xi - r(x - x_0))}(x)$$
$$= (p^{(1)} + p^{(2)})\psi_{(\xi)}(x) + r(p^{(1)} - p^{(2)})\psi_{(x - x_0)}(x).$$

Let us first find $p^{(i)}$ such that $q \equiv 1$ in Theorem 5.1.5. Then we want to have $q = p^{(1)} + p^{(2)} = 1$ on $\partial D$. Hence on $\partial D$ we find

$$p^{(1)}(x) = \frac{1}{2} - \frac{\psi_{(\xi)}(x)}{2r\psi_{(x - x_0)}(x)}, \qquad p^{(2)}(x) = \frac{1}{2} + \frac{\psi_{(\xi)}(x)}{2r\psi_{(x - x_0)}(x)}.$$

Observe that since $\psi$ is concave, for $x \in \partial D$, we have

(5.2.2)
$$\psi_{(x - x_0)}(x) = -\psi(x_0) + \frac{1}{2}\psi_{(x - x_0)(x - x_0)}(\theta) \le -\psi(x_0) < 0,$$

where $\theta$ is a point between $x$ and $x_0$ on the straight line passing through these points.

The assumption about the smoothness of $D$ allows us to continue $p^{(i)}$ from the boundary inside $D$ in such a way that for thus obtained functions for which we keep the same notation we have $p^{(i)} \in C^2(\bar{D})$. Furthermore, owing to (5.2.2) we can do the continuation so that

$$|p^{(i)}|_{1,D} \le N|\xi|r^{-1}\psi^{-2}(x_0), \qquad |p^{(i)}|_{2,D} \le N|\xi|r^{-1}\psi^{-3}(x_0).$$

Now

$$|p^{(i)}| \le N(1 + |\xi|r^{-1}\psi^{-1}(x_0)), \qquad E_{x_0}|\xi_\tau^{(i)}| \le N(|\xi| + r\psi^{1/2}(x_0)),$$

so that (5.1.8) implies

$$|u_{(\xi)}(x_0)| \le N|g|_{1,D}(1 + |\xi|r^{-1}\psi^{-1}(x_0))(|\xi| + r\psi^{1/2}(x_0))$$
$$+ N|g|_{1,D}|\xi|r^{-1}\psi^{-3}(x_0)(|\xi| + r\psi^{1/2}(x_0)).$$

By taking $r = |\xi|/\sqrt{\psi(x_0)}$ we come to

(5.2.3)                    $|u_{(\xi)}(x_0)| \le N|g|_{1,D}|\xi|\psi^{-5/2}(x_0).$

We have obtained a result which is slightly weaker than our claim by using a "natural" $q$. Now we improve it. To satisfy (5.1.7), which is (5.2.1) in our case, this time we choose

(5.2.4)  $p^{(1)}(x) = -r\psi_{(x-x_0)}(x) + \psi_{(\xi)}(x), \quad p^{(2)}(x) = -r\psi_{(x-x_0)}(x) - \psi_{(\xi)}(x)$

for all $x \in \bar{D}$. Obviously, $p^{(i)} \in C^2(\bar{D})$. Then due to (5.2.2) on $\partial D$ we have

(5.2.5)         $q := p^{(1)} + p^{(2)} = -2r\psi_{(x-x_0)}(x) \ge 2r\psi(x_0).$

Using the smoothness of $D$ we continue $q$ from the boundary inside $D$ and get a function $q(x), x \in \bar{D}$, such that $q \in C^1(\bar{D})$. Again (5.2.2) allows us to do the continuation so that

$$|q^{-1}|_{1,D} \le Nr^{-1}\psi^{-2}(x_0), \quad |g/q|_{1,D} \le Nr^{-1}|g|_{1,D}\psi^{-2}(x_0).$$

Finally,

$$|p^{(i)}|_{2,D} \le N(r + |\xi|), \quad E_{x_0}|\xi_\tau^{(i)}| \le N(|\xi| + r\psi^{1/2}(x_0)),$$

which along with (5.1.8) implies

$$|u_{(\xi)}(x_0)| \le Nr^{-1}|g|_{1,D}\psi^{-2}(x_0)(r + |\xi|)(|\xi| + r\psi^{1/2}(x_0)).$$

By taking $r = |\xi|$ we come to $|u_{(\xi)}(x_0)| \le N|g|_{1,D}|\xi|\psi^{-2}(x_0)$. The theorem is proved.                                                                        □

REMARK 5.2.2. The above proof can be adjusted to cover the case when $0 \in D$ and we want to only estimate $u_{(\xi)}(0)$ under the additional assumption that there is a real-valued bounded function $\lambda(x)$ such that $\sigma_{(x)}(x) = \lambda(x)\sigma(x)$. If $\lambda \equiv 0$ this condition means that $\sigma$ is constant in radial direction.

Indeed, if we define quasiderivatives $\xi_t^{(i)}$, $i = 1, 2$, on the basis of (3.2.1) again with $\pi \equiv 0$ and $P \equiv 0$ and $r_t^{(i)} = (-1)^i r(1 - \lambda(x_t))$, then (3.2.1) becomes

$$\xi_t^{(i)} = \xi + \int_0^t [\sigma_{(\xi_s^{(i)})}(x_s) + (-1)^i r(1 - \lambda(x_s))\sigma(x_s)] \, dw_s,$$

which has the solution $\xi_t^{(i)} = \eta_t + (-1)^i r x_t$, where $\eta_t$ is the solution of the above equation with $r = 0$. However, to satisfy (5.1.7) which is

$$p^{(1)}\psi_{(\eta_\tau - r x_\tau)} + p^{(2)}\psi_{(\eta_\tau + r x_\tau)} = 0$$

we have to take functions $p^{(i)}$ depending not only on $x$ but also on the extra coordinate $\eta$. We will see later that such extra coordinates also appear if $b \ne 0$. Of course, one also has to impose a condition on $\sigma$ guaranteeing that $\eta_{t \wedge \tau}$ is uniformly integrable.

## 5.3. A general result in time dependent situation

If in the situation of Section 5.2 we allow constant $b \neq 0$, then only using time change based quasiderivatives seems to be not enough. Indeed, in that case for constant $r$ we have

$$\xi_t = \xi + r\sigma^k w_t^k + 2rbt = \xi + 2r(x_t - x_0) - r\sigma^k w_t^k .$$

In Section 5.2 by taking two different $r$ and using linear combinations of $\xi + 2r(x_t - x_0)$ with weights $p^{(i)}$ we were able to steer the linear combination into the tangent plane to $\partial D$ at $x_\tau$. However this time we also have to deal with terms like $\sigma^k w_\tau^k$. We decided to make them disappear in the end by involving the parameters $\pi$ associated with change of measure. These parameters contribute terms $\sigma^k \tau$ to the first quasiderivatives. By taking linear combination of such terms with weights $w^k/\tau$ we balance out the new terms. Important point to notice here is that we need to use weights depending not only on $x_\tau$ but also $w_\tau$ and $\tau$ and the weights now are not bounded and do not have bounded derivatives.

This is why we need the following generalization of Lemma 5.1.1. For $s \in \mathbb{R}$ and $x \in \mathbb{R}^d$ consider the following equation

(5.3.1) $\qquad dx_t = \sigma^k(s + t, x_t)\, dw_t^k + b(s + t, x_t)\, dt, \quad t \geq 0, x_0 = x ,$

where $w_t^k$ and Borel functions $\sigma^k$, $k = 1, \dots, d_1$, and $b$ have the same meaning as before. We assume that there is a finite constant $K$ dominating $\sigma, b$ and their first order derivatives in $x$. Then equation (5.3.1) has a unique solution $x_t(s, x)$. Similarly to $E_x$ we introduce $E_{s,x}$ for expectations of certain quantities defined in terms of $x_t(s, x)$.

In this section $D$ is a possibly unbounded domain in $\mathbb{R}^d$,

$$\tau = \tau(s, x) = \tau_D(s, x)$$

is the first exit time of $x_t(s, x)$ from $D$. We assume that for all $s \in \mathbb{R}$ and $x \in D$

$$E_{s,x}\tau \leq K .$$

Define

(5.3.2) $\qquad Lu(s, x) = \partial u(s, x)/\partial s + a^{ij}(s, x)u_{x^i x^j}(s, x) + b^i(s, x)u_{x^i}(s, x) .$

As usual in the parabolic setting, for a function $g$ given in $\bar{Q}_T$ with $Q_T := (T, \infty) \times D$ we denote

$$[g]_{1,\bar{Q}_T} = \sup_{t \geq T}[g(t, \cdot)]_{1,\bar{D}} + \sup_{x \in D}[g(\cdot, x)]_{1/2,[T,\infty)}, \quad |g|_{1,\bar{Q}_T} = |g|_{0,\bar{Q}_T} + [g]_{1,\bar{Q}_T} .$$

LEMMA 5.3.1. *Let $y \in \partial D$, $s \in \mathbb{R}$, $n \in \mathbb{R}^d$, $|n| = 1$, $\varepsilon_0 > 0$. Assume that the straight segment $\Lambda := \{y + \varepsilon n : 0 < \varepsilon \leq \varepsilon_0\}$ lies in $D$ and*

(5.3.3)                         $$E_{s,y+\varepsilon n}\tau \leq K\varepsilon \quad \forall 0 < \varepsilon \leq \varepsilon_0 .$$

*Let $p$ be a continuous function on $\bar{Q}_s$ such that the derivatives $p_t$, $p_x$, $p_{xx}$ are continuous in $Q_s$ and $p(s, \cdot) \in C^1(\bar{\Lambda})$. Assume that there are (finite) constants*

$$|\sigma^* p_x|_{s,y} \quad and \quad |Lp|_{s,y}$$

*such that for $0 < \varepsilon \leq \varepsilon_0$ we have*

(5.3.4)
$$E_{s,y+\varepsilon n} \int_0^\tau |\sigma^* p_x(s+t, x_t)|^2 \, dt \leq \varepsilon |\sigma^* p_x|_{s,y}^2 ,$$
$$E_{s,y+\varepsilon n} \int_0^\tau |Lp(s+t, x_t)| \, dt \leq \varepsilon |Lp|_{s,y} .$$

*Let $g$ be a function with $|g|_{1,\bar{Q}_s} < \infty$. Introduce*

(5.3.5)            $$u(s, x) = E_{s,x} g(s + \tau, x_\tau), \quad v(s, x) = E_{s,x} p g(s + \tau, x_\tau)$$

*and assume that $u(s, \cdot), v(s, \cdot) \in C^1(\bar{\Lambda})$. Then*

(5.3.6)
$$|v_{(n)}(s, y) - p u_{(n)}(s, y)| \leq |p_{(n)} g(s, y)|$$
$$+ N[g]_{1,\bar{Q}_s} |\sigma^* p_x|_{s,y} + |g|_{0,\bar{Q}_s} |Lp|_{s,y} ,$$

*where $N$ depends only on $K$.*

PROOF. By repeating the proof of Lemma 5.1.1 with obvious changes, instead of (5.1.5) we find that

$$v(s, y + \varepsilon n) \leq p u(s, y + \varepsilon n) + \varepsilon (N[g]_{1,\bar{Q}_s} |\sigma^* p_x|_{s,y} + |g|_{0,Q_s} |Lp|_{s,y}) .$$

After that repeating the rest of the proof of Lemma 5.1.1 immediately leads to (5.3.6) and proves our lemma.                                                  □

REMARK 5.3.2. Obviously one can take $K^{1/2}|\sigma^* p_x|_{0,\bar{Q}_s}$ and $K|Lp|_{0,\bar{Q}_s}$ in place of $K|\sigma^* p_x|_{s,y}$ and $K|Lp|_{s,y}$, respectively, in (5.3.4).

The following lemma is proved in exactly the same way as Corollary 5.1.4.

LEMMA 5.3.3.

(i) *Let $\partial D$ be once continuously differentiable, $\xi \in \mathbb{R}^d$, $x_0 \in D$, $s_0 \in \mathbb{R}$.*

(ii) *For each $s > s_0$ and $y \in \partial D$ let there exist an $\varepsilon_0 > 0$ such that condition (5.3.3) is fulfilled with $n = n(y)$ being the inward unit normal to $\partial D$ at $y$.*

(iii) *Let $p$ be a continuous function which is defined in $(s_0, \infty) \times \bar{D}$ such that the derivatives $p_t$, $p_x$, $p_{xx}$ are continuous in $(s_0, \infty) \times D$ and $p_x$ is continuous in $(s_0, \infty) \times \bar{D}$. Also assume that there are Borel functions*

$$|\sigma^* p_x|_{s,y} \quad \text{and} \quad |Lp|_{s,y}$$

*defined on $(s_0, \infty) \times \partial D$ such that for any $s > s_0$, $y \in \partial D$, and $0 < \varepsilon \leq \varepsilon_0$ condition (5.3.4) is satisfied.*

(iv) *Let $g$ be a function with $|g|_{1,\bar{Q}_{s_0}} < \infty$ such that the functions $u$ and $v$ introduced in (5.3.5) are well defined in*

$$\bar{Q}_{s_0} \setminus (\{s_0\} \times \partial D)$$

*and are continuously differentiable in $x$ on this set.*

(v) *Let an $\mathbb{R}^d$-valued process $\xi_t$ be a quasiderivative of $x_t(s_0, x)$ at $x_0$ with adjoint process $\xi_t^0$ and $\xi_0 = \xi$. Assume that the process*

$$v_{x^i}(x_{t \wedge \tau}) \xi_{t \wedge \tau}^i + v(x_{t \wedge \tau}) \xi_{t \wedge \tau}^0$$

*is uniformly integrable, where $x_t = x_t(s_0, x_0)$ and $\tau = \tau(s_0, x_0)$.*
*Then*

(5.3.7)
$$|v_{(\xi)}(s_0, x_0) - E_{s_0, x_0} \xi_\tau^0 pg(s_0 + \tau, x_\tau) - E_{s_0, x_0} pu_{(\xi_\tau)}(s_0 + \tau, x_\tau)|$$
$$\leq 2|g|_{0, \bar{Q}_{s_0}} E_{s_0, x_0} |p_x(s_0 + \tau, x_\tau)| \, |\xi_\tau|$$
$$+ N|g|_{1, \bar{Q}_{s_0}} E_{s_0, x_0} |(n(x_\tau), \xi_\tau)| (|\sigma^* p_x|_{s_0 + \tau, x_\tau} + |Lp|_{s_0 + \tau, x_\tau}).$$

Now follows the main result of this section. Its proof is obtained on the basis of Lemma 5.3.3 in the same way as Theorem 5.1.5 is derived from Corollary 5.1.4.

THEOREM 5.3.4. *Let assumptions (i) and (ii) of Lemma 5.3.3 be satisfied.* *Suppose that for $k = 1, \ldots, m$ we are given some objects*

$$p^{(k)}, \quad |\sigma^* p_x^{(k)}|_{s,y}, \quad |Lp^{(k)}|_{s,y}$$

*having the same meaning as in Lemma 5.3.3 and satisfying assumption (iii) of Lemma 5.3.3 for each $k$. Let $q$ be a function bounded away from zero with $|q|_{1,\bar{Q}_{s_0}} < \infty$ and such that*

$$q(t, x) = \sum_{k=1}^{m} p^{(k)}(t, x) \quad \text{on} \quad (s_0, \infty) \times \partial D.$$

*Let g be a function with* $|g|_{1,\bar{Q}_{s_0}} < \infty$ *such that, for* $\bar{g} := g/q$, *the functions*

$$u(s, x) = E_{s,x} g(s + \tau, x_\tau), \quad \bar{u}(s, x) = E_{s,x} \bar{g}(s + \tau, x_\tau),$$
$$v^{(k)}(s, x) = E_{s,x} p^{(k)} \bar{g}(s + \tau, x_\tau), \quad k = 1, \dots, m$$

*are well defined in*

$$\bar{Q}_{s_0} \setminus (\{s_0\} \times \partial D)$$

*and are continuously differentiable in x on this set.*

*For* $k = 1, \dots, m$, *let* $\xi_t^{(k)}$ *be quasiderivatives of* $x_t(s_0, x)$ *at point* $x_0$ *with adjoint processes* $\xi_t^{(0k)}$ *and* $\xi_0^{(k)} = \xi$. *Assume that, for* $k = 1, \dots, m$, *the processes (there is no summation in k below)*

(5.3.8) $$v_{x^i}^{(k)}(s_0 + t \wedge \tau, x_{t \wedge \tau}) \xi_{t \wedge \tau}^{(k)i} + v^{(k)}(s_0 + t \wedge \tau, x_{t \wedge \tau}) \xi_{t \wedge \tau}^{(0k)},$$

*are uniformly integrable, where* $x_t = x_t(s_0, x_0)$ *and* $\tau = \tau(s_0, x_0)$.
   *Then*

$$\left| u_{(\xi)}(s_0, x_0) - E_{s_0,x_0} \left[ \sum_{k=1}^m p^{(k)} \bar{g}(s_0 + \tau, x_\tau) \xi_\tau^{(0k)} + \bar{u}_{(\bar{\xi}_\tau)}(s_0 + \tau, x_\tau) \right] \right|$$

$$\leq N|\bar{g}|_{1,D} \sum_{k=1}^m E_{s_0,x_0} |\xi_\tau^{(k)}| \left( |p_x^{(k)}(s_0 + \tau, x_\tau)| + |\sigma^* p_x^{(k)}|_{s_0+\tau,x_\tau} + |Lp^{(k)}|_{s_0+\tau,x_\tau} \right),$$

*where*

$$\bar{\xi}_\tau = \sum_{k=1}^m p^{(k)}(s_0 + \tau, x_\tau) \xi_\tau^{(k)}.$$

In the next section we need the following.

COROLLARY 5.3.5. *If*

(5.3.9) $$\bar{\xi}_\tau \perp n(x_\tau) \quad (a.s.),$$

*where* $x_t = x_t(s_0, x_0)$, $\tau = \tau(s_0, x_0)$, *then*

$$|u_{(\xi)}(s_0, x_0)| \leq \left| E_{s_0,x_0} \sum_{k=1}^m [p^{(k)} \bar{g}(s_0 + \tau, x_\tau) \xi_\tau^{(0k)} + p^{(k)} \bar{g}_{(\bar{\xi}_\tau^{(k)})}(s_0 + \tau, x_\tau)] \right|$$

(5.3.10)
$$+ N|\bar{g}|_{1,D} \sum_{k=1}^m E_{s_0,x_0} |\xi_\tau^{(k)}| \left( |p_x^{(k)}(s_0 + \tau, x_\tau)| \right.$$

$$+ |\sigma^* p_x^{(k)}|_{s_0+\tau,x_\tau} + |Lp^{(k)}|_{s_0+\tau,x_\tau} \right).$$

## 5.4. General case of constant coefficients in uniformly convex domains

Here as in Section 5.2 we consider equation (0.0.1) with constant $\sigma$ and $b$ but without assuming that $b$ is zero. Assume that

$$\operatorname{tr} \sigma \sigma^* + |b| = 1 .$$

Again, we take $D \in C^3$ as a uniformly convex domain in $\mathbb{R}^d$.

It is trivial that $E_x \tau \leq N$ where $N$ depends only on the diameter of $D$. However, in contrast with Section 5.2 now there is no guarantee that $E_x \tau$ goes to zero as $D \ni x \to \partial D$ not slower than $\operatorname{dist}(x, \partial D)$. This happens because $|b|$ can be large in comparison with $\operatorname{tr} \sigma \sigma^*$. If $\operatorname{tr} \sigma \sigma^* = 0$, then at some points on $\partial D$ the function $E_x \tau$ will not even go to zero. Assuming that $\operatorname{tr} \sigma \sigma^* \geq \varepsilon^2 > 0$ does not change much the situation. For instance, it is not hard to show that, if $D \subset \mathbb{R}^2$ is the unit disk $\{|x| \leq 1\}$, $d_1 = 1$, $\sigma^1$ is the first basis vector times $\varepsilon$, $b = (0, 1 - \varepsilon^2)$, then for any $\alpha \in (0, 1)$ there is an $\varepsilon > 0$ such that

$$E_{(0,y)} \tau \geq (y + 1)^\alpha$$

for all $y \in (-1, 0)$ sufficiently close to $-1$. This example can be treated by our methods just by enlarging the domain for $y > y_0$ so as to have a rather sharp curvature of the boundary near the south pole if we are only interested in the estimates at points $(x, y)$ with $y > y_0$. Here $b$ is orthogonal to the direction in which the diffusion is moving. On the other hand the situation in which $d \geq 1$ and $b = \sigma^k c_k$ with some constants $c_k$ can be easily reduced to to the one with $b = 0$ just by using Girsanov's theorem.

However, we do not know how to treat more general situations and therefore we assume that there is a function $\phi$ such that

$$\phi \in C^2(\bar{D}), \quad \phi = 0 \quad \text{on} \quad \partial D, \quad \phi > 0 \quad \text{and} \quad L\phi \leq -1 \quad \text{in} \quad D .$$

Then by Itô's formula

$$E_x \tau \leq \phi(x) \leq N \operatorname{dist}(x, \partial D) ,$$

where $N$ is a constant. In that case our estimates are independent of any further relations between $\sigma$ and $b$.

THEOREM 5.4.1. *Under the above assumptions take a* $g \in C^1(\bar{D})$ *and introduce* $u(x) = E_x g(x_\tau)$. *Then $u$ is locally Lipschitz continuous in $D$ and (a.e.)*

$$(5.4.1) \qquad |u_x(x)| \leq N |g|_{1,D} \operatorname{dist}(x, \partial D)^{-11/2} ,$$

*where $N$ depends only on $D$ and $\phi$.*

PROOF. As in the proof of Theorem 5.2.1, without losing generality we assume that $a$ is uniformly nondegenerate and $g$ and $D$ are infinitely differentiable. Consider the two-component process $z_t = (x_t, y_t)$ given by

$$(5.4.2) \qquad dx_t = \sigma\, dw_t + b\, dt, \quad dy_t = dw_t,$$

in the domain $D' = D \times \mathbb{R}^{d_1}$. Notice that the first exit time of $(x_t, y_t)$ from $D'$ is just the first exit time of $x_t$ from $D$, so that in the notation of Section 5.3

$$u(x) = E_{s,x,y}\, g(x_\tau).$$

We represent the quasiderivatives of $z_t$ as $\zeta_t = (\xi_t, \eta_t)$, where $\xi_t \in \mathbb{R}^d$ and $\eta_t \in \mathbb{R}^{d_1}$ and since we are only interested in the derivatives of $u$ in $x$, we take

$$\zeta_0 = (\xi, 0), \quad x_0 \in D, \quad y_0 = 0, \quad s_0 = 0.$$

First as before we take two time change related quasiderivatives. Let $\zeta_t^{0i} \equiv 0$, $i = 1, 2$,

$$d\xi_t^{(1)} = \sigma^k\, dw_t^k + 2b\, dt, \quad d\eta_t^{(1)} = dw_t,$$
$$d\xi_t^{(2)} = -\sigma^k\, dw_t^k - 2b\, dt, \quad d\eta_t^{(1)} = -dw_t.$$

Then

$$\xi_t^{(1)} = \xi + 2(x_t - x_0) - \bar{\xi}_t, \quad \eta_t^{(1)} = w_t$$
$$\xi_t^{(2)} = \xi - 2(x_t - x_0) + \bar{\xi}_t, \quad \eta_t^{(2)} = -w_t,$$

where

$$\bar{\xi}_t = \int_0^t \sigma\, dw_s = \sigma^k w_t^k.$$

Take a convex $\psi \in C^3(\bar{D})$ such that

$$\psi = 0 \quad \text{on} \quad \partial D, \quad \psi > 0 \quad \text{in} \quad D,$$

so that a vector $\kappa$ is tangential to $\partial D$ at a point $y \in \partial D$ if and only if $\psi_{(\kappa)}(y) = 0$.

Take the same $p^{(i)}(t, x, y) = p^{(i)}(x)$ as in (5.2.4) but with 2 in place of $r$. Then with

$$t = \tau, \quad x = x_\tau = x_0 + \sigma w_\tau + b\tau, \quad y = y_\tau = w_\tau, \quad \bar{\xi} = \bar{\xi}_\tau$$

we have

$$(5.4.3) \qquad \begin{aligned} & p^{(1)}(t, x, y)\psi_{(\xi + 2(x - x_0) - \bar{\xi})}(x) + p^{(2)}(t, x, y)\psi_{(\xi - 2(x - x_0) + \bar{\xi})}(x) \\ & = -2\psi_{(\bar{\xi})}(x)\psi_{(\xi)}(x) = -2y^k \psi_{(\sigma^k)}(x)\psi_{(\xi)}(x). \end{aligned}$$

To make this vanish we use measure change related quasiderivatives and for $k = 1, \ldots, d_1$ introduce $\xi_t^{(01k)} = -\xi_t^{(02k)} = w_t^k$,

$$\xi_t^{(1k)} = \xi + \int_0^t \sigma^k \, dt = \xi + \sigma^k t, \quad \eta_t^{(1,k)} = t,$$

$$\xi_t^{(2k)} = \xi - \int_0^t \sigma^k \, dt = \xi - \sigma^k t, \quad \eta_t^{(2,k)} = -t,$$

$$p^{(1,k)}(t, x, y) := \psi_{(\xi)}(x) \frac{y^k}{t} =: -p^{(2,k)}(t, x, y).$$

Then with the same $t, x, y$ as in (5.4.3)

$$\sum_{k=1}^{d_1} p^{(1,k)}(t, x, y) \psi_{(\zeta_t^{(1k)})}(x) + \sum_{k=1}^{d_1} p^{(2,k)}(t, x, y) \psi_{(\zeta_t^{(2k)})}(x)$$

$$= \sum_{k=1}^{d_1} p^{(1,k)}(t, x, y) (\psi_{(\xi_t^{(1k)})}(x) - \psi_{(\xi_t^{(2k)})}(x))$$

$$= 2t \sum_{k=1}^{d_1} p^{(1,k)}(t, x, y) \psi_{(\sigma^k)}(x) = 2y^k \psi_{(\sigma^k)}(x) \psi_{(\xi)}(x).$$

It follows that

$$\sum_{k=1}^{d_1} p^{(1,k)}(t, x, y) \psi_{(\zeta_t^{(1k)})}(x) + \sum_{k=1}^{d_1} p^{(2,k)}(t, x, y) \psi_{(\zeta_t^{(2k)})}(x)$$

$$+ p^{(1)}(t, x, y) \psi_{(\zeta_t^{(1)})}(x) + p^{(2)}(t, x, y) \psi_{(\zeta_t^{(1)})}(x) = 0$$

and the vector

$$\sum_{k=1}^{d_1} p^{(1,k)}(\tau, x_\tau, y_\tau) \zeta_\tau^{(1k)} + \sum_{k=1}^{d_1} p^{(2,k)}(\tau, x_\tau, y_\tau) \zeta_\tau^{(2k)}$$

$$+ p^{(1)}(\tau, x_\tau, y_\tau) \zeta_\tau^{(1)} + p^{(2)}(\tau, x_\tau, y_\tau) \zeta_\tau^{(2)}$$

is tangential to $\partial D'$ at $(x_\tau, y_\tau)$.

Obviously, the same function $q$ from (5.2.5) equals the sum of all $p^{(i)}$, $p^{(jk)}$ over $i, j = 1, 2$, $k = 1, \ldots, d_1$ in $(0, \infty) \times \partial D'$ (notice that the value $t = 0$ is excluded). $\square$

To check other assumptions of Theorem 5.3.4 we need the following.

LEMMA 5.4.2. *Let $\alpha \geq 0$, $\beta \geq 1$ be some constants. Then there is a constant $N$ depending only on $D$ and $\alpha$, $\beta$ such that, for $s \geq 0$ and $x \in D$,*

(5.4.4)
$$E_{s,x,y} \int_0^\tau \frac{1}{(s+t)^\alpha} \, dt \leq N \frac{1}{s^\alpha} \psi(x),$$

(5.4.5)
$$E_{s,x,y} \int_0^\tau \frac{|y_t|^\beta}{(s+t)^\alpha} \, dt \leq N \frac{|y|^\beta + 1}{s^\alpha} \psi(x).$$

(5.4.6)
$$E_x \frac{1}{(s+\tau)^\alpha} \leq N \frac{1}{s^\alpha + \psi^{2\alpha}(x)},$$

(5.4.7)
$$E_x \frac{|w_\tau|^\beta}{(s+\tau)^\alpha} \leq N \frac{\psi^{1/2}(x)}{s^\alpha + \psi^{2\alpha}(x)}.$$

PROOF. Estimate (5.4.4) follows from the fact that $(s+t)^{-\alpha} \leq s^{-\alpha}$. We use the same observation and also notice that since $y_t = y + w_t$ is a martingale, $|y_t|^\beta$ is a submartingale. Then the left-hand side of (5.4.5) times $s^\alpha$ turns out to be less than

$$\int_0^\infty E_x |y + w_t|^\beta I_{\tau>t} \, dt \leq \int_0^\infty E_x |y + w_\tau|^\beta I_{\tau>t} \, dt$$

$$= E_x |y + w_\tau|^\beta \tau \leq |y|^\beta E_x \tau + (E_x |w_\tau|^{2\beta})^{1/2} (E_x \tau^2)^{1/2}.$$

Here by the Burkholder-Davis-Gundy inequalities and Remark 5.1.3

$$E_x |w_\tau|^{2\beta} \leq N E_x \tau^\beta \leq N E_x \tau.$$

This yields (5.4.5).

While proving (5.4.6) it suffices to concentrate on $s = 0$ since $(s+\tau)^{-\alpha}$ is less than both $s^{-\alpha}$ and $\tau^{-\alpha}$ and in addition $a^{-\alpha} \wedge b^{-\alpha} \leq 2^\alpha (a+b)^{-\alpha}$.

Denote $\rho = \text{dist}(x, \partial D)$ and observe that

$$\rho \leq |x_\tau - x_0| \leq |w_\tau| + \tau.$$

Therefore, by exponential estimates, for any $\varepsilon \in (0, \varepsilon_0)$, where $\varepsilon_0 := \rho/2$, we have
$$P_x(\tau < \varepsilon) \leq P_x(\sup_{t \leq \varepsilon} |w_t| \geq \rho - \varepsilon)$$

$$\leq P_x(\sup_{t \leq \varepsilon} |w_t| \geq \rho/2) \leq N e^{-\rho^2/(8\varepsilon)}.$$

Hence,

$$E_x \tau^{-\alpha} = \int_0^\infty P_x(\tau < t^{-1/\alpha}) \, dt \leq \varepsilon_0^{-\alpha} + N \int_{\varepsilon_0^{-\alpha}}^\infty e^{-\rho^2 t^{1/\alpha}/8} \, dt$$

$$\leq N\rho^{-\alpha} + N \int_0^\infty e^{-\rho^2 t^{1/\alpha}/8} \, dt = N\rho^{-\alpha} + N\rho^{-2\alpha} \leq N\rho^{-2\alpha}.$$

This proves (5.4.6). To prove (5.4.7) it suffices to use Cauchy's inequality along with the fact that $E_x |w_\tau|^{2\beta} \leq N E_x \tau^\beta \leq N\psi(x)$. The lemma is proved.  □

Lemma 5.4.2 implies, in particular, that for $s \geq 0$ the function

$$v^{(ik)}(s, x, y) = E_{s,x,y} p^{(ik)}(s + \tau, x_\tau, y_\tau)\bar{g}(x_\tau)$$

are well defined and (with $N$ depending on $x_0$)

$$|v^{(ik)}(s, x, y)| \leq N|\bar{g}|_{0,D}|\xi|E_x \frac{|y + w_\tau|}{s + \tau} \leq N\frac{|y| + \psi^{1/2}(x)}{s + \psi^2(x)}.$$

To prove that $v^{(ik)}$ are smooth in $\bar{Q}_0 \setminus (\{0\} \times \partial D')$, where $Q_0 = (0, \infty) \times D'$, it is convenient to assume that $\sigma = \sqrt{2a}$. Obviously this assumption does not restrict generality and allows us to write $y_t = y_t(s, x, y) = w_t + y = \sigma^{-1}(x_t - x - bt) + y$. Hence,

$$\begin{aligned} v^{(ik)}(s, x, y) &= E_x p^{(ik)}(s + \tau, x_\tau, \sigma^{-1}(x_\tau - b\tau))\bar{g}(x_\tau) \\ &\quad + (y - \sigma^{-1}x)^k E_x p^{(ik)}(s + \tau, x_\tau, e_k)\bar{g}(x_\tau) \\ &=: \bar{v}^{(ik)}(s, x) + (y - \sigma^{-1}x)^k \tilde{v}^{(ik)}(s, x), \end{aligned}$$

where $e_k$ is the $k$th basis vector. This formula expresses $v^{(ik)}$ through the solutions $\bar{v}^{(ik)}$ and $\tilde{v}^{(ik)}$ of the equation $\partial v/\partial t + Lv = 0$ in $[0, \infty) \times D$. Since the equation is nondegenerate, local regularity results (see, for instance, Chapter 4 in [16]) show that $v^{(ik)}$ is indeed infinitely differentiable in $\bar{Q}_0 \setminus (\{0\} \times \partial D')$ and

$$(5.4.8) \quad |v^{(ik)}_{x,y}(s, x, y)| \leq N\frac{|y|+1}{(s + \psi^2(x))\psi(x)} \leq N\frac{|y|+1}{\psi^3(x)} \leq N(1 + |y|^2 + \psi^{-6}(x)).$$

Furthermore, if $s_0 > 0$ and $x_0 \in \partial D$, then in the intersection of

$$(s_0/2, 3s_0/2) \times \{x : |x - x_0| < \sqrt{s_0/2}\} \quad \text{with} \quad (0, \infty) \times \partial D$$

the $C^3$-norms of the boundary data of $\bar{v}^{(ik)}$ and $\tilde{v}^{(ik)}$ are bounded by a constant times $1 + s_0^{-3}$. From boundary and interior regularity results for parabolic equations we now get that, for each $s > 0$, $|\bar{v}^{(ik)}_x(s, x)|$ and $|\tilde{v}^{(ik)}_x(s, x)|$ are dominated by a constant times $1 + s^{-7/2}$. It follows that

$$|v^{(ik)}_{x,y}(s, x, y)| \leq N(1 + |y|^2 + s^{-7}),$$

which along with (5.4.8) yields

$$\begin{aligned} |v^{(ik)}_{x,y}(s, x, y)| &\leq N(1 + |y|^2 + s^{-7} \wedge \psi^{-6}(x)) \\ &\leq N(1 + |y|^2) + N\frac{1}{s^7 + \psi^6(x)}. \end{aligned}$$

Now fix an $x \in D$, define

$$D_1 = \{y \in D : \psi(y) > \psi(x)/2\}$$

and let $\gamma$ be the first exit time of $x_t = x_t(x)$ from $D_1$. Notice that, if $t \le \gamma$, then $\psi(x_t) > \psi(x)/2$. However, if $t \ge \gamma$, then $t^7 \ge \gamma^7$. Since by Lemma 5.4.2 applied to $D_1$ we have

$$E_x \gamma^{-7} \le \psi^{-14}(x) \,,$$

the above argument shows that, for any $x \in D$,

$$E_x \sup_{t \le \tau} \frac{1}{t^7 + \psi^6(x_t)} \le E_x \sup_{t \le \gamma} \frac{1}{t^7 + \psi^6(x_t)} + E_x \sup_{\gamma \le t \le \tau} \frac{1}{t^7 + \psi^6(x_t)} < \infty \,.$$

One can also easily estimate

$$E_x \sup_{t \le \tau} |\xi_t^{(ik)}|^2$$

and make other necessary computations to see that the requirement in Theorem 5.3.4 about processes corresponding to (5.3.8) is satisfied in our particular situation.

To apply Theorem 5.3.4 we also need to check condition (iii) in Lemma 5.3.3 for $p^{(i)}$, $p^{(jk)}$. Denote by $p$ any one of these functions and notice that for the operator $L$ as in (5.3.2) associated with $z_t$ we have

$$|\sigma^* p_z(t, x, y)| \le N(1 + |\xi|) + N|\xi| \left( 1 + \frac{|y| + 1}{t} \right) \,,$$

$$|Lp(t, x, y)| \le N(1 + |\xi|) + N|\xi| \left( 1 + \frac{|y| + 1}{t} + \frac{|y| + 1}{t^2} \right) \,.$$

Lemma 5.4.2 shows that as $|\sigma^* p_z|_{s,x,y}$ and $|Lp|_{s,x,y}$ for $p = p^{(i)}$, $p^{(jk)}$ and $(s, x, y) \in (0, \infty) \times D'$ we can take

$$N + N|\xi|(|y| + 1)(1 + t^{-2})$$

and then

$$(5.4.9) \quad E_{0,x_0,0}(|p_z(\tau, z_\tau)|^2 + |\sigma^* p_z|_{\tau, x_\tau, y_\tau}^2 + |Lp|_{\tau, x_\tau, y_\tau}^2) \le N(1 + |\xi|^2 \psi^{-8}(x_0)) \,.$$

We can finally use (5.3.10). Observe that the summation there is to include all terms corresponding to $p^{(i)}$, $p^{(jk)}$. First we are dealing with $p^{(jk)}$. We have

$$I_1 := \sum_{i=1}^{2} \sum_{k=1}^{d_1} p^{(ik)} \bar{g}(\tau, x_\tau) \xi_\tau^{(0ik)} = 2\bar{g}(\tau, x_\tau) \psi_{(\xi)}(x_\tau) |w_\tau|^2 / \tau \,,$$

$$I_2 := \sum_{i=1}^{2} \sum_{k=1}^{d_1} p^{(ik)} \bar{g}_{(\xi_\tau^{(ik)})}(\tau, x_\tau) = 2\bar{g}_{(\sigma k)} w_\tau^k \psi_{(\xi)}(x_\tau) \,.$$

By Lemma 5.4.2

$$E_{0,x_0,0}(|I_1| + |I_2|) \leq N|\xi|\psi^{-1-2+1/2}(x_0) + N|\xi|\psi^{-2+1/2}(x_0) \,.$$

Owing to (5.4.9) and the estimate

$$E_{0,x_0,0}|\xi_\tau^{(ik)}|^2 \leq N(|\xi|^2 + \psi(x_0)) \,,$$

we see that the last term on the right in (5.3.10) corresponding to $p^{(jk)}$ is less than

$$N\psi^{-2}(x_0)(1 + |\xi|\psi^{-4}(x_0))(|\xi| + \psi^{1/2}(x_0)) \,.$$

Similarly and somewhat easier one estimates the terms corresponding to $p^{(1)}$, $p^{(2)}$ and then according to (5.3.10) we conclude

$$|u_{(\xi)}(x_0)| \leq N|\xi|\psi^{-5/2}(x_0) + N\psi^{-2}(x_0)(1 + |\xi|\psi^{-4}(x_0))(|\xi| + \psi^{1/2}(x_0)) \,.$$

We substitute here $\psi^{1/2}(x_0)\xi/|\xi|$ in place of $\xi$ and obtain (5.4.1) thus proving the theorem.

# Bibliography

[1]   Y. A. ALKHUTOV, *The behavior of solutions of parabolic second-order equations in noncylindrical domains*, Dokl. Akad. Nauk **345** n. 5 (1995), 583–585 (in Russian).

[2]   J. M. BISMUT, *Martingales, the Malliavin calculus and hypoellipticity under general Hörmander's conditions*, Z. Warsch. Verw. Gebiete **56** (1981), 469–505.

[3]   H. DONG, *About Smoothness of Solutions of the Heat Equations in Closed Smooth Space-time Domains*, submitted to Comm. Pure Appl. Math.

[4]   I. I. GIKHMAN, *Certain differential equations with random functions*, Ukraïn. Mat. Zh. **2** n. 4 (1950), 37–63; **3** n. 3 (1951), 317–339 (in Russian).

[5]   M. I. FREIDLIN, *On the smoothness of solutions of degenerate elliptic equations*, Izv. Akad. Nauk SSSR, ser. math. **32** n. 6 (1968), 1391–1413 in Russian; English translation: Math. USSR-Izv. **2** n. 6 (1968), 1337–1359.

[6]   J. J. KOHN – L. NIRENBERG, *Degenerate elliptic-parabolic equations of second order*, Comm. Pure Appl. Math. **20** n. 4 (1967), 551–585.

[7]   V. A. KONDRAT'EV, *Boundary value problems for parabolic equations in closed regions*, Trudy Moskov. Mat. Obšč. **15** (1966), 400–451 (in Russian); English translation: 450–504 in Transactions of the Moscow Mathematical Society for the Year 1966. Translation from the Russian prepared jointly by the American Mathematical Society and the London Mathematical Society, Amer. Math. Soc., Providence, R.I. 1967.

[8]   N. V. KRYLOV, *Smoothness of the value function for a controlled diffusion process in a domain*, Izv. Akad. Nauk SSSR, ser. mat. **53** n. 1 (1989), 66–96 (in Russian); English translation: Math. USSR Izvestija **34** n. 1 (1990), 65–96.

[9]    N. V. KRYLOV, *On first quasiderivatives of solutions of Itô's stochastic equations*, Izv. Akad. Nauk SSSR, ser. mat. **56** n. 2 (1992), 398–426; English translation in Russian Acad. Sci. Izv. Math. **40** n. 2 (1992), 377–403.

[10]   N. V. KRYLOV, *Quasiderivatives for solutions of Itô's stochastic equations and their applications*, pp. 1-44 in Stochastic Analysis and Related Topics (Proc. of the Fourth Oslo-Silivri Workshop on Stoch. Analysis, Oslo, July 1992), ed. T. Lindström et al, Stochastic Monographs, Vol. 8, Gordon an Breach Sci. Publishers, Switzerland etc., 1993.

[11]   N. V. KRYLOV, *A theorem on degenerate elliptic Bellman equations in bounded domains*, Differ. Integral Equ. Appl. **8** n. 5 (1995), 961–980.

[12]   N. V. KRYLOV, "Introduction to the theory of diffusion processes", Amer. Math. Soc., Providence, RI, 1995.

[13]   N. V. KRYLOV, *Adapting some ideas from stochastic control theory to studying the heat equation in closed smooth domains*, Appl. Math. Optim. **46** n. 2/3 (2002), 231–261.

[14]   N. V. KRYLOV, *Quasiderivatives and interior smoothness of harmonic functions associated with degenerate diffusion processes*, to appear in Electronic Journal of Probability.

[15]   N. V. KRYLOV – B. L. ROZOVSKY, "Stochastic evolution equations", Itogi nauki i tekhniki, Vol. 14, VINITI, Moscow, 1979, 71-146 (in Russian); English translation in J. Soviet Math. **16** n. 4 (1981), 1233–1277.

[16]   G. LIEBERMAN, "Second order parabolic differential equations", World Scientific, Singapore-New Jersey-London-Hong Kong, 1996.

[17]   O. A. OLEINIK – E. V. RADKEVICH, "Second order equations with non-negative characteristic form", Itogi Nauki, Mat. Analis, 1969, VINITI, Moscow, 1971 (in Russian); English translation: Amer. Math. Soc., Plenum Press, Providence R.I., 1973.

[18]   A. THALMAIER, *Gradient estimates for harmonic functions on regular domains in Riemannian manifolds*, J. Funct. Anal. **155** (1998), 109–124.

# PUBBLICAZIONI DELLA CLASSE DI SCIENZE
## DELLA SCUOLA NORMALE SUPERIORE
### QUADERNI

1. DE GIORGI E., COLOMBINI F., PICCININI L.C.: *Frontiere orientate di misura minima e questioni collegate.*
2. MIRANDA C.: *Su alcuni problemi di geometria differenziale in grande per gli ovaloidi.*
3. PRODI G., AMBROSETTI A.: *Analisi non lineare.*
4. MIRANDA C.: *Problemi in analisi funzionale* (ristampa).
5. TODOROV I.T., MINTCHEV M., PETKOVA V.B.: *Conformal Invariance in Quantum Field Theory.*
6. ANDREOTTI A., NACINOVICH M.: *Analytic Convexity and the Principle of Phragmén-Lindelöf.*
7. CAMPANATO S.: *Sistemi ellittici in forma divergenza. Regolarità all'interno.*
8. TOPICS IN FUNCTIONAL ANALYSIS: Contributors: F. STROCCHI, E. ZARANTONELLO, E. DE GIORGI, G. DAL MASO, L. MODICA.
9. LETTA G.: *Martingales et intégration stochastique.*
10. OLD AND NEW PROBLEMS IN FUNDAMENTAL PHYSICS: Meeting in honour of GIAN CARLO WICK.
11. INTERACTION OF RADIATION WITH MATTER: A Volume in honour of ADRIANO GOZZINI.
12. MÉTIVIER M.: *Stochastic Partial Differential Equations in Infinite Dimensional Spaces.*
13. SYMMETRY IN NATURE: A Volume in honour of LUIGI A. RADICATI DI BROZOLO.
14. NONLINEAR ANALYSIS: A Tribute in honour of GIOVANNI PRODI.
15. LAURENT-THIÉBAUT C., LEITERER J.: *Andreotti-Grauert Theory on Real Hypersurfaces.*
16. ZABCZYK J.: *Chance and Decision. Stochastic Control in Discrete Time.*
17. EKELAND I.: *Exterior Differential Calculus and Applications to Economic Theory.*
18. ELECTRONS AND PHOTONS IN SOLIDS: A Volume in honour of FRANCO BASSANI.
19. ZABCZYK J.: *Topics in Stochastic Processes.*
20. TOUZI N.: *Stochastic Control Problems, Viscosity Solutions and Application to Finance.*

### CATTEDRA GALILEIANA

1. LIONS P.L.: *On Euler Equations and Statistical Physics.*
2. BJÖRK T.: *A Geometric View of the Term Structure of Interest Rates.*
3. DELBAEN F.: *Coherent Risk Measures.*

### LEZIONI LAGRANGE

1. VOISIN C.: *Variations of Hodge Structure of Calabi-Yau Threefolds.*

### LEZIONI FERMIANE

1. THOM R.: *Modèles mathématiques de la morphogénèse.*
2. AGMON S.: *Spectral Properties of Schrödinger Operators and Scattering Theory.*
3. ATIYAH M.F.: *Geometry of Yang-Mills Fields.*
4. KAC M.: *Integration in Function Spaces and Some of Its Applications.*
5. MOSER J.: *Integrable Hamiltonian Systems and Spectral Theory.*
6. KATO T.: *Abstract Differential Equations and Nonlinear Mixed Problems.*
7. FLEMING W.H.: *Controlled Markov Processes and Viscosity Solution of Nonlinear Evolution Equations.*
8. ARNOLD V.I.: *The Theory of Singularities and Its Applications.*
9. OSTRIKER J.P.: *Development of Larger-Scale Structure in the Universe.*

10. NOVIKOV S.P.: *Solitons and Geometry.*
11. CAFFARELLI L.A.: *The Obstacle Problem.*
12. CHEEGER J.: *Degeneration of Riemannian metrics under Ricci curvature bounds.*

## PUBBLICAZIONI DEL CENTRO DI RICERCA MATEMATICA ENNIO DE GIORGI

1. DYNAMICAL SYSTEMS. Part I: *Hamiltonian Systems and Celestial Mechanics.*
2. DYNAMICAL SYSTEMS. Part II: *Topological, Geometrical and Ergodic Properties of Dynamics.*
3. MATEMATICA, CULTURA E SOCIETÀ 2003.
4. RICORDANDO FRANCO CONTI.
5. N. KRYLOV: *Probabilistic Methods of Investigating Interior smoothness of Harmonic Functions Associated with Degenerate Elliptic Operators.*

## ALTRE PUBBLICAZIONI

*Proceedings of the Symposium on FRONTIER PROBLEMS IN HIGH ENERGY PHYSICS* Pisa, June 1976

*Proceedings of International Conferences on SEVERAL COMPLEX VARIABLES,* Cortona, June 1976 and July 1977

*Raccolta degli scritti dedicati a JEAN LERAY apparsi sugli Annali della Scuola Normale Superiore di Pisa*

*Raccolta degli scritti dedicati a HANS LEWY apparsi sugli Annali della Scuola Normale Superiore di Pisa*

*Indice degli articoli apparsi nelle Serie I, II e III degli Annali della Scuola Normale Superiore di Pisa* (dal 1871 al 1973)

*Indice degli articoli apparsi nella Serie IV degli Annali della Scuola Normale Superiore di Pisa* (dal 1974 al 1990)

ANDREOTTI A.: *SELECTA vol. I, Geometria algebrica.*

ANDREOTTI A.: *SELECTA vol. II, Analisi complessa, Tomo I e II.*

ANDREOTTI A.: *SELECTA vol. III, Complessi di operatori differenziali.*

Fotocomposizione "CompoMat" Loc. Braccone, 02040 Configni (RI), Italy
Finito di stampare per conto della "CompoMat" dalla Nuova Grafica 86 nel novembre 2004